WAR ZONES TO WISDOM

THE PRICE OF PEACE

A true story of survival, struggle, and the courage
to become more than what the world expected.

JACOB ROWE

Warzones to Wisdom: The Price of Peace

A true story of survival, struggle, and the courage to become more than the world expected. From small town poverty to the battlegrounds of Afghanistan, this individual's journey reveals that survival is only the beginning.

Warzones to Wisdom: The Price of Peace is a gripping, unfiltered account of growing up surrounded by violence and hardship, then stepping into war with a sense of duty. Walking away with scars, strength, and a deeper sense of purpose.

After facing the chaos of combat and the silence that follows, the author rebuilt his life from the ground up, turning pain into power and setbacks into steppingstones. Through real estate, coaching, counseling and entrepreneurship, Jacob Rowe carved a path to freedom, not just financially, but mentally and emotionally.

This story isn't just about surviving war, it is about finding peace, redefining success, and becoming more than what the world ever expected. Told with raw honesty and hard-won wisdom, this memoir is a testament to resilience, faith, and the kind of transformation that can only come through fire.

CONTENTS

About the Author

Jacob Rowe grew up in Sandusky, Ohio, the middle child of five children. His early life was marked by hardship, two older brothers trapped in cycles of addiction and incarceration, alongside a father who battled paranoid schizophrenia. Home was rarely safe or steady, but it built in him a quiet resilience that would shape everything that followed.

At just nineteen years old, Jacob joined the military in search of something more stable and meaningful. He went on to serve in Afghanistan, where the realities of combat deeply transformed him. Before leaving the military, he found new direction through real estate, coaching, and entrepreneurship, creating a future he was never supposed to have.

Jacob wrote *Warzones to Wisdom: The Price of Peace* for anyone who has ever been judged by where they came from. His story is proof that we are more than our circumstances, more than our past, and more than what the world says we are.

More than anything, he wrote this book as a legacy for his kids, to show them that no matter where you start, you always have the power to choose where you go next.

Why I'm Telling This Story Now

I never planned to write a book. I'm not a celebrity and I didn't grow up craving the spotlight. I sure didn't think I had a story anyone would care to hear. But life has a way of changing your perspective—especially when you become a father, a leader, and a man who's made it through things that were supposed to break you. And this is what I've learned: *silence helps no one.*

This book is not a "war story" or a way to glorify war. This book is for anyone who's ever been told they wouldn't make it. For the kid growing up in violence. For the soldier carrying invisible wounds. For the person trying to change while being haunted by who they used to be. This story is for you.

Most of all, I'm writing this for my kids. I want them to know the truth, not just the version of me they see now, but everything it took to become this man I am today. I want them to understand that I've been beaten down but never broken. That I've seen darkness but I've also stood back up. Again, and again.

During my deployment in Afghanistan, I sustained multiple traumatic brain injuries. Since then, memory loss has become part of my life, sometimes subtle, sometimes sharp, but always present. That is why I'm writing this now. While I still remember. While the memories are still mine to share. This book has been written to the best of my recollection. Memory is not a perfect thing and nobody remembers everything perfectly. However, this life, these lessons, they matter. And I do not want them to get lost.

I have walked through poverty, violence, war, and hardship. But I've also walked through healing, growth, purpose, and peace. I have seen what happens when you just keep going, even when the world expects you to quit. And I have learned that peace is not free. It costs something. Something inside yourself. Peace requires the death of the bits and pieces of who you were so you can be more. But it is worth every scar. If this story gives even one person hope… then it was worth it.

CHAPTER 1

Sandusky-Surviving the Unseen War at Home

I grew up in Sandusky, Ohio—middle child of five. Two older brothers. Two younger sisters. A full house, sure. But truthfully? It often felt like we were just trying to survive under the same roof. We lived in a single-family home with 3 bedrooms. My mom had a room, and my sister Hannah had a bedroom. My brothers and I shared a room upstairs until my grandma moved in with us. After that, my brothers and I all shared the living room as our bedroom. The house we lived in was on Megs Street. If you are from the area, you know this was not considered the nicest area. The Police station was less than a mile down the road but that did not mean they would show up quickly if you called. Our backyard butted up to the lake and a row of boat houses. The owners of the boat houses were not thrilled to have us there and often tried to have our home condemned by the court so we would be forced to move out and the house would be torn down. They considered us an eye sore. The town was pretty beat down in general. Lots of old abandoned factory buildings and drugs riddled the area. All the money went to the main strip where all the tourists were drawn to due to Cedar Point. Cedar Point is one of the largest amusement parks, not just in Ohio but also the United States, that people would travel to from all over the country. Cedar Point brought in all sorts of people and money, but the money never seemed to come back to the community. The schools were old brick buildings and most of them, from when I was younger, are closed now. Sandusky still had its upsides, I guess. But from where I sat, I never really got to experience that part.

My mom worked nights as an ER nurse, doing everything she could to hold our world together. She was strong, but bone-tired. My dad was paranoid schizophrenic. Unmedicated, unpredictable, and far too often, violent. He took it out on my mom and sometimes on me when I tried to get in the way. He ruled the house with fear. We lived on edge, never knowing what would set him off next. Even as a kid, I felt it was my job to protect my little sisters from the worst of it. I took that on before I even knew what protection really meant. But the truth is, long before they were born, one night shaped the way I saw the world.

I was about four years old. My parents were screaming at each other in the kitchen. I can't even remember what sparked it. But I remember the sound of my mom crying like it was yesterday. I remember seeing her on the floor, my dad standing over her, hitting her. She screamed at me to call the cops. I stood there frozen, terrified. But I knew I had to do something. I picked up the phone and started to dial 911. I didn't get far before my dad ripped the phone from my hands and smashed it across my face. I hit the ground hard. He looked at me and said, "This is how you handle a woman who doesn't do what she's told." That moment is carved into my memory. At four years old, I didn't understand everything that was happening, but I knew that was wrong. And right then, I made a silent promise to myself: *No one in my life will ever feel that kind of fear—not because of me, or because of anyone I can protect them from.*

That night lit a fire in me. It established a line I knew I would never cross. Even before I had the words for it, I didn't want to repeat the pain I came from. I wanted to be a protector.

As I got older, I saw the cycle play out around me. My brothers fell into violence, drugs, and prison. The streets didn't care if they lived or died. And then one day, the cycle came for me.

My oldest brother, who'd also been abused by our dad, turned his anger on me. He came at me with everything he had. And in that moment, I saw it clearly: *this is how it passes down.* This went on for years. Drunken outbursts of rage and fists. Fights that had nothing to do with me and yet, there I sat. Holes in walls, broken doors, screaming, and the cops showing up at our door regularly.

I had two choices—follow what I'd been taught or fight like hell to become something else. I didn't get it all right. I made mistakes. I carried the weight of what I'd seen and felt. But that fire never went out. Even in those dark times, I believed there had to be another way.

During this time, we all were members of a church, Lighthouse Baptist Church. There were two specific men that brought our family in. To this day I still have no idea what brought them to us, so I see it as an act of God. There was no family connection or friendship before this. Just two men who were going door to door to recruit others to assist them in building this church. They were asking for help in the community to clear out an old building. From this, we agreed to help. My Mom, Dad, brothers, and myself all pitched in time and effort on this church. This was the only time in my life I can remember doing anything together that has a lasting impression. I can only remember one man vividly; his name was Brother Ben. I think he saw what home life was for us and tried his best to be there for us. Brother Ben had his own struggles in life, though. Even though he was a selfless and godly man, he was filled with sorrow. Brother Ben spent time with us, taught us how to read the Bible, sacrificed his own free time to take us places and look after us when my parents couldn't be around. He organized nerf gun wars, took us to church, helped at the house if we needed him to stop by. He never treated us the way that other people seen us. Like we were trash or not worth the time. Brother Ben had his own family that he couldn't protect from drugs, addiction, and I believe that was the tipping point for him. He took his own life when I was about six years old and I believe that was truly the turning point in my brothers lives and when they stepped into the streets. They believed the friends they made in the streets were their real family. This was really when I started to separate from them and try to find my own way. I lost the connection with my brothers in order to live a different life.

Sandusky gave me my beginning. But more than that, it gave me the reason to leave—and a vow to never become the man I was taught to be.

CHAPTER 2

Fremont—A New Kind of Quiet

When I was 9 years old, my mom left my dad. She took me and two of my siblings to live with her and her boyfriend in Fremont, a town of about fifteen thousand people and a half hour's drive from Sandusky. Fremont wasn't a clean break from the past—but it was a turning point.

When we moved in with my mom's boyfriend, Troy, everything shifted. It wasn't just a change of address; it was a change in atmosphere. The streets were quieter. The houses stood a little sturdier. People waved when they walked by, like it was normal to be seen and acknowledged. The energy was calmer, more predictable. It felt like we'd stepped into someone else's version of life, a version where safety didn't have to be earned every day. But even with all that, I didn't feel like I belonged.

Fremont was polished in a way I wasn't. The people there moved differently. They smiled more, laughed louder. Their clothes looked fresh, like they came from stores I didn't walk into. Their problems seemed... softer. Even though they weren't cruel, I could feel them sizing me up—measuring me against whatever invisible checklist they used to decide whether someone fit in. It wasn't hate. It was subtler than that. A kind of quiet, polite doubt. The kind you only recognize when you've spent your life being underestimated. I didn't talk like them. I didn't carry myself like them. I didn't know how to be "at ease." I was rough around the edges in a town full of rounded corners.

At first, it was just me, my younger sister, Hannah, and my brother, Zach, who made the move. Zach was 11 or 12 years old at the time but his time in Fremont didn't last long. Not even a year in, he got caught stealing from Troy. Maybe it was habit, maybe it was rebellion. Maybe Fremont made him feel even more out of place than I did. Either way, it blew up. We found out quickly that Zachary did not want this new life. He did not want to get along with Troy and he wanted to go back to Sandusky. Troy was a real 'rough around the edge" guy. He was very foreign to how we lived prior to this. Troy was a Marine in his youth. He was more structured than we were. He had family dinners growing up and lived more of a country life than we were used to. We did not know it then, but Troy would grow into a husband for my mother, and a father to my sisters and I. You cannot force a kid to grow into a family that they did not want to be a part of. Troy and Zachary could not see eye to eye on things, so my mother gave Zach the option to go back to Sandusky. This did not mean going back to live with my biological father. There was a restraining order in place, he was court ordered to not come in contact with my mom or us. The violence eventually came out and my dad spent years in and out of prison. So, my mom told Zach that he could move back to Sandusky with my grandma if that's what he wanted. He packed up and moved back to Sandusky as fast as he could. After that, the energy in the house shifted. The air got tenser, the lines got clearer.

Then, one day, my mom sat me down. "Do you want to go back too?" she asked. I thought about it. I missed my friends. I missed the rhythm of a place I understood. I missed how sports in Sandusky made me feel—like I had worth, like I had a role. But I couldn't go. As much as I missed those things, I couldn't leave my sister, Hannah. She was still little. Still figuring things out. And in that house, in that town, she needed someone in her corner. I couldn't trust anyone else to protect her the way I would. That moment became one of the first real tests of character I can remember. A crossroads. I could choose what was comfortable. Or I could choose what was right. Right isn't always loud. It's not always rewarded. It's not always obvious. Sometimes, right means staying put when everything in you wants to run. So, I stayed.

I ended up making friends due to my athleticism but there was only one person that was a true friend to me at that time. His name was Robert, but we all called him Bob. Bob was an asshole—crass, mean, aggressive, and angry— but so was I. We got along well. After yelling at each other and fighting with each other, we decided that we were glad to meet someone else willing to stand their ground. We were very different but we both chose to be who we wanted to be and not what others told us to be. It was a respect for strong character. We didn't do drugs or follow the parties. We played sports, chilled, and worked out. I would go on long runs and Bob would just play games until I was done. There were several people who would have enjoyed my company, but it was hard for me to relate to them or even find common ground. I had a lot of acquaintances, but Bob was the only friend I kept from a young age until now.

Eventually, I knew I needed to earn my own income so I would not have to rely on anyone or ask for support. Asking for things has always felt very uncomfortable to me. No one taught me that I shouldn't ask or need help, life just taught me there is a difference between what you need and what you want. If it was a want, I needed to earn it myself. I started working as soon as I could. My first job was in the back kitchen at a little restaurant called Chud's Bar and Grill in Fremont. It was hot, loud, fast-paced, exactly what I needed. The restaurant was busy enough for me to work but not big enough to keep me from being able to have a life outside of work. It was a small brick building on the corner of the street. Red brick, white grout, with a big Clock on the front. In here I could just be an employee. I wasn't a bad kid that didn't fit the environment. It was the kind of motion I could control. All that needed to be seen was my work ethic and ability to learn and take direction. No one knew me because of my last name or actions of my family that I had no control over. No one was reading the newspaper and asking me what was going on in my family. I also picked up work on a nearby farm, baling hay, shoveling, doing anything that needed doing. I didn't care what the job was, I just wanted to earn my way. To contribute.

There was power in it. After the life I had seen unfold at a young age, I realized something: The people who want more, who expect something for nothing, who use words like "deserve," do not understand that they already have more than they should. I have seen this firsthand. I did not want money without work; on top of that, how I conducted my work mattered. I did not want to be an athlete to become rich or famous. I didn't need to be remembered by the masses. I just wanted to have impact on the life and position I was facing at that moment. To have a presence and name I could build for myself. I relished the idea that I could wake up every day and choose the work I wanted to do and the effort I wanted to give. My impact would be great even if I was just cleaning a toilet. Because I was doing it. It was my responsibility, and I wanted to take extreme ownership of my actions and my presence.

For the first time, I wasn't hustling. I wasn't stealing. I wasn't scheming or surviving off someone else's pity. I was earning money. I was sweating for something. And even if nobody else saw it, I did. That mattered to me. I was so used to a life where the people around me got together to plan a robbery or to sell drugs for quick cash. It was easier and faster. The problem was that it led to an empty and antagonizing life. There is no peace for the man who builds his empire on short cuts and preying on the weak.

When I wasn't working, in school, or helping my sister, I was running or working out at the gym. Sometimes after a late shift—when most kids were out partying or doing whatever reckless things teenagers do—I'd lace up my shoes and take off. No music. No destination. Just me, my breath, and the road. From one county line to another. Midnight to 2 a.m. Just running.

The rhythm of my feet hitting the pavement was the only thing in my life that felt consistent. That silence out there... That was therapy. That was survival. It was how I made sense of everything that didn't make sense. The anger, the loneliness, the fear all had somewhere to go in those miles. Every step was proof that I was still moving. Still choosing my own direction. The weather did not matter. In the winter months I ran in Timberland boots. They were warm, strong, and held up to the test of 10,000,000 steps.

Running kept my body disciplined and my mind clear. It kept me from becoming the kind of man I swore I'd never be. I didn't party. I didn't drink. I didn't chase distractions. I stayed focused. Though it made me feel like more of an outsider, I knew I needed that distance.

The truth is those runs kept me alive. They gave me something to do when my mind wanted to wander. Self-pity gets you nowhere. I hate pity and, quite frankly, the word "trauma." Pity is such an empty emotion. Pity isn't care or empathy. It is just a way for someone to look down on you and pretend to care when they are really self-reflecting or trying to change the subject from whatever they felt pity about, to something that made them more Comfortable. When a person can experience empathy and compassion, that is different. They can feel your pain or sorrow or even joy. They do not have to understand exactly what it is like to be you, but they can place themselves in the image you are painting and see how much it would affect them, and they can draw from that emotion to know how to listen and be still. They do not have to make it better or make you feel better, they just need to be willing to be there in that moment with you . Overall, I had nothing to be sad about. Nothing to whine about. I just had shit to do if I did not want to keep living the same life I came from.

Even though Fremont was healthier, it never really felt like *my* home. It felt like a home for my mom. For Troy, who officially became my stepfather in 2009 and whom I love as a father. And for my two sisters. I was grateful for it—but I also felt like a guest in someone else's life. Like I was allowed to be in the picture, but only at the edge of the frame. I was a memory connected to a past that did no favors. My sisters seemed more like a fresh start for Mom and Troy. My younger sister Gi, who was Troy's daughter and not my biological sister, was quickly accepted into the family when Troy got full custody of her. We welcomed her and there was never even a discussion. My Mom and Troy leaped at the chance to bring her into the home. This wasn't a bad thing in my eyes. This was great. I was happy it could be this way. I didn't want to get in the way of this new family. I felt on the outside of it all. My mom, Troy, Hannah, and Gi all seemed like a family, and I was just there to fill in or help whenever someone needed it. I knew my sisters still needed emotional and mental support. So did mom and Troy at times. I could pull myself out of the situation to attend to all sides long enough to make sure everyone had a voice and an outlet. When my mom would have nervous breakdowns about my older brothers not doing well or about money. When Troy would need an outlet because he was stretched too thin to be everything for everyone while having no time to nourish his own life. For my sisters when they could not understand why the fighting and arguing in the house was so loud and aggressive. It was everyone going through their own versions of war with themselves, so I had to do what I could to show up for them in the ways I could. I'm sure it wasn't enough, but it was all I had to give, myself, my time, my presence. Still, it was better than what I came from and sometimes, better is enough.

In my senior year of high school, Zachary, who still lived in Sandusky, was thrown out of a third-story window. The investigators on the case couldn't completely piece together what happened, but the aftermath was obvious: Zachary hit his head and ended up with a fractured skull and cerebral hemorrhaging. The doctors were not certain he would survive, and even if he did, they were not sure what his life would look like after that.

I sat in the hospital with my mom, watching her crumble. Troy was there, trying to be supportive, but in those moments there's nothing to say and nothing to do—except stay present. No one took the easy way out. Everyone could have kept living their life. They could have just gone back to school or work and let the hospital do its job but that never happened. We came together to support Zachary in his time of need. Even if there were things we were all upset with him for… We still showed up. Troy and Zach were not on good terms at this point. But Troy set all that aside and did what he could to support. Troy was not our father by blood, but he chose to be there when it counted even though we were not his responsibility. I think this speaks volume about a man's character in general. In my eyes, Troy was a father to Zach as well, even if neither one of them will ever say it out loud. Troy made mistakes as we all do, but he has earned forgiveness in all the ways people may have viewed him as coming up short. It's easy to make the wrong call as a parent… But knowing they will come for you in a time of need is really the main point. My Mom, my sisters Hannah and Gi, Troy, and myself all showed up for Zach when he needed us the most.

Throughout the night and into the next morning, the doctor informed us that the swelling was going down and the bleeding was subsiding. He didn't know why; he just knew that Zachary was pulling through.

We were relieved but still scared of what waited on the other side. When the tubes were removed from Zach's mouth so he could breathe on his own and speak, his voice was different, quiet and raspy. His vocal cords were traumatized, and he would never speak the same again. But at least he could speak.

In the next coming months, we all spent as much time as we could at the hospital with him, trying to be there for him in every way we could, helping him walk, helping him into the bathroom. We helped him learn to speak clearly again, and how to read. Our whole lives became about Zachary. And that was exactly what needed to happen. When someone you care about endures something like that, you do all you can. I hope people would show up for me in the ways that we showed up for Zach. I offered prayer and time. I offered compassion and love.

Even still, I'm sure it wasn't enough. I'm sure he had several quiet nights and angry feelings. I'm sure he felt lonely and disconnected. I'm sure he needed more than we could give. Zach's injury was another example of why I had to leave Sandusky. The place we called home did not care about us. There was no love lost in the community. We were known because my oldest brother was dealing drugs. Known for the violence that flowed in and out of the house. For all the nights the cops were at our door looking to arrest my brothers or my father. I was sick of going places and being known for the actions of other people (like my own family members) instead of my own. I was sick of law-enforcement, viewing me in a negative light whenever they found out what my last name was. I wanted to be anonymous—easy to overlook if not completely invisible—so I could move through life quietly. I wanted to help more than I hurt and to give back more than I took.

As I got older, something began to settle in me. A realization: I couldn't control where I came from—but I could control what I did with it. My choices mattered. How I treated people mattered. The way I carried myself, the standard I held, the boundaries I set, all of it became part of my identity. I didn't need anyone to tell me I was doing the right thing. I just had to live it. I could be an example instead of a warning. I could lead by how I moved, not by what I said. I could show my sisters, and later, my children, that the cycle could stop here. Being a victim of your environment isn't a badge of honor; it's a curse of false belief and failed teachings.

Toward the end of high school, that belief became something more. I became a father. My first son, Ezekiel, was born in my senior year—and everything changed the second I held him. That moment wasn't loud either. It didn't come with fireworks. But something in me shifted in a way that was permanent. His mother, Kayla, and I seen different ways to approach life. We took different paths and did not end up together for the journey. Regardless of this, Kayla has always been a kind and loving mother to Ezekiel. She does what she believes is right and would do anything for our son. Kayla also has wonderful parents which gives Zeke 2 sets of loving and admiring Grandparents. When Ezekiel was born, I made him a promise—without saying a word. You'll never feel the fear I grew up with. You'll know love. You'll never wonder if you matter.

That promise became my new compass. Every job I took. Every early morning. Every decision to walk away from drama, from easy money, from old habits—that promise led me. This is when I truly started to set boundaries in my life. No more bailing my brothers out of jail. No more letters to the judge. No more giving money to grown people to pay their own bills. For too long in my life I have carried the weight of other mistakes. I was fine with it because I knew I would be OK. I knew I could handle it. But now, it wasn't about me anymore.

Fremont wasn't the end of my pain. But it was the start of my purpose. Distance truly gives clarity. Take some time in your life to zoom out. See the full picture. And sometimes, that's the bigger victory.

CHAPTER 3

The Military Decision

After high school, life didn't slow down—it hit the gas. There was no gap year, no breather, no easing into adulthood. I went straight into full-time work at the Whirlpool factory, grinding out twelve-hour shifts on the line. It was hot, loud, soul-numbing work—same movement, same noise, same sweat, day after day. But it paid. And I had a child to care for. At this time, I was paying child support to Kayla because Zeke lived with her and her parents. They offered me a place in their home very generously so I could be closer to Zeke and help more often with the baby. At this time in my life, I just did not know how to accept the kindness and I felt uncomfortable living under another man's roof. They offered with good intention, but I was not in a place to accept. I felt that I had to work and figure it out as I went.

Luxury didn't exist in my world. Not financial luxury. Not time. Every hour counted. At night, after work, I'd head to school to become a firefighter. I didn't have time for parties. No late-night hangs, no distractions. Just work, school, parenting—on repeat. The exhaustion was real, but I didn't complain. I had a purpose. I wasn't drifting. I was building.

Becoming a firefighter felt like progress. It would mean health insurance for Ezekiel. It would mean structure. Predictability. It would mean I could be present in a way my father never was. I was stretched thin—but I was moving with intention. The problem came when I learned how hard it was to get a full-time position in the department. Openings do not come often—probably because most firefighters love their job. I was a volunteer for a while, but it wasn't enough. I couldn't provide my family with the life I wanted, and something still felt incomplete.

As I looked around at the life I was grinding out, I started to feel a weight that didn't come from exhaustion, but from stagnation. The people around me didn't seem to care much about what they were doing—or why they were doing it. Their eyes told the story. Eyes that said, "This is all there is." That scared me more than failure. Because I knew if I stayed there too long, I'd start to believe it too. And I didn't survive everything I'd been through just to end up in neutral.

Ever since I was young, I'd carried two goals in my chest like a secret pact: I wanted to be a firefighter. And I wanted to be a soldier. At the time, I didn't have the language for why—only a feeling. But looking back now, it was simple: I wanted to be a protector. I wanted to be strong enough to stand between danger and the people who couldn't fight for themselves. That instinct had been with me since I was four years old, when I stood frozen with a phone in my hand and blood on my lip. It lived in my bones.

Volunteering as a firefighter scratched the surface of that instinct. It gave me a glimpse. But something deeper inside me was still unsettled. I knew I had another promise to keep, one that had nothing to do with titles and everything to do with transformation.

The more I thought about the military, the louder the call got. Not just as a job. Not as a way out. But as a way up. A way to break the ceiling I felt closing in. A way to test the man I was becoming, to find out if I had what it took, not just to survive, but to serve something bigger.

Because the truth is, I didn't want the life that was in front of me. I wanted to see life differently than how I was taught to view it. I wanted something that mattered more than myself. I wanted to do something with my life that my children could be proud of. As their father, it is my responsibility to show them how to live rather than just stand next to them and tell them.

The only other dream that came close was to open my own gym someday, a space where people could come to suffer on purpose, to build strength like the way I did through discipline, sweat, and the transformation of pain into power. But I knew that dream would take time. First, I had to finish the promises I made to that younger version of me. I had to prove to him we were still on mission.

College didn't connect for me. I did the classes. I passed the tests. But my heart wasn't in it. It felt like a placeholder, not a path. A waiting room for people still figuring it out. And I already knew, I wasn't built to wait. I was built to move.

The military felt like motion. By then, things back at my mom's house had shifted. My mom had gone back to school herself—at night, just like I had. She was getting her master's degree. She was thriving. My sisters were older, stronger, finding their own voices. The household didn't feel broken anymore. It felt stable. And for the first time, it felt like no one needed rescuing.

Ezekiel was living with his mom, and the rest of my family was on the right path. I felt that gave me the freedom to leave. To become something my child could look up to. To show him different ways to live and grow and that the end of one thing could be the beginning of something else. I also wanted to be able to provide for him more than a paycheck-to-paycheck life. This way he could have health care and access to different schooling options for college.

So, I enlisted. I didn't want an easy desk job. I wanted a role that mattered. I didn't want to just exist in the military; I wanted something sharp, active, dangerous. Something that would stretch me and shape me and make me dangerous in the right way. I wanted to be dependable. I wanted to be forged. I chose the role of combat engineer because it was both warrior and builder. Not just boots on the ground but hands in the dirt. It meant demolitions, construction, breaching obstacles, and clearing routes under fire. I could be useful in a firefight and valuable when the smoke cleared. It gave me something I could carry with me, on and off the battlefield. I didn't join just to serve my country. I joined to serve the mission I'd been living for since childhood: Protect those who couldn't protect themselves. Carry the weight. I joined to step fully into that role. No more waiting. No more wondering. It was time to move.

CHAPTER 4

Boot Camp–Finding Order in Chaos

I expected boot camp to be extreme. Everything I'd heard or seen growing up painted it as a relentless blur of screaming, sweating, and being broken down to your core, only to be rebuilt into a warrior. Sure, there was plenty of yelling. There were early mornings, brutal workouts, and moments designed to push you past your limits. But compared to my youth? Boot camp felt… calm. Relaxing.

For the first time in my life, there was order, rules, structure, and clear expectations. It was loud, but it made sense. There was a rhythm to it. When you've lived in a house where fear ruled the day and violence could show up without warning, structure, no matter how hard, feels like peace.

Before I left, my stepfather pulled me aside. "Be invisible," he said. "Don't stand out. If they don't notice you, they won't mess with you." That was his advice and what helped him in the Marine Corps. At the time, that advice felt smart. I didn't go to basic to be a star. I didn't want attention. I wanted to finish, get through it, and get to the fight. I tried to lay low. To blend in. Head down, mouth shut. That strategy didn't last long. What I hadn't realized was that I'd already been preparing for boot camp for years. The gym had been my sanctuary. While other people partied, I trained. While they slept in, I ran. While they wasted time, I focused. The weight room taught me consistency, suffering, and repetition. It taught me how to show up, even when nobody else was watching.

When the physical demands of boot camp hit, I was more ready than I knew. Where others struggled, I thrived. I ran faster. Pushed harder. Recovered quicker. I wasn't the loudest. I wasn't the biggest. I was the most prepared. While some recruits were trying to survive, I was sharpening my skills. Because I knew what was coming, it was 2011 and we were at war. I hadn't enlisted for adventure. I hadn't come for some "GI Joe" fantasy. I just wanted to build a life to be proud of. I didn't want to waste my time. I am not a fucking statistic. Life is a choice. The harder the choice, the better the outcome. It has almost been a certainty in my life. Always choose the harder path to walk and God will show you it was the right path in the end. Comfortability is the death of all dreams. I hoped to become lethal, efficient, and useful. Because I knew this: The life expectancy of a combat engineer in a warzone isn't generous.

Within a few weeks, I was named platoon guide, top recruit, responsible for leading the others. It's not a title that carries over into real Army life, but in that moment, it was the first real leadership test I'd ever faced. I didn't want the role. I didn't need it, but I accepted it. Not because I wanted to be in charge, but because I took the mission seriously.

I earned respect the same way I always had, by doing the work. Not barking orders. No ego involved. I simply showed up. I executed, listened, and I outworked ever excuse. And that made people follow. Leadership doesn't come from volume; it comes from consistency.

Boot camp taught me a lot. Honestly, most of the lessons weren't new. They just had names now. Words for what I'd already lived through growing up: *Stay sharp. Stay ready. Carry your weight. Don't complain. Be useful.*

The difference was this time I wasn't doing it alone. I was surrounded by other men trying to earn something. Men who had come from their own struggles. And for the first time, I felt like I was part of something bigger. Something I could trust. By the end of basic, I had broken physical fitness records. I graduated in the Hall of Fame for my class. That had never been the goal, but it was confirmation. Proof that the way I'd been living, disciplined, focused, intentional. That I wasn't just surviving. I was preparing.

When I got my orders for Fort Drum, New York, I knew deployment was next. My unit already had a combat rotation on the calendar. Afghanistan was waiting and I was ready to meet it.

CHAPTER 5

Afghanistan—Combat, Survival, Transformation

My friend Bynoe and I were transferred to a Special Troops Battalion to support efforts for deployment. Preparation for the deployment started months before we left. It helps you train and prepare when you know you are slotted for an actual combat deployment. I attended classes on dismount techniques, bomb disposal, Russian munitions, route clearance, room and village clearing tactics. Some of it felt theoretical, like we were still roleplaying war. But deep down, we knew this training was about keeping each other alive.

What struck me most during those training months was the quiet shift in mindset. Every briefing and drill became heavier. Every piece of gear became personal. I wasn't just packing equipment. I was packing what might be my last set of memories. The laughter during breaks got louder, almost desperate, like we were trying to store up joy before it was taken from us.

We all took leave for Christmas and spent time with family before we headed out. Early January 2013 was when we were slotted to depart. That was right around the time I found out my girlfriend and future wife, Caitlyn, was pregnant. This complicated things in our lives quite a bit because we had only been dating for a few short months at the time, but it was worth all the complications. I remember watching her sleep the night before I left, her hand on her stomach, and wondering if I'd ever meet our child. We couldn't change the situation we were in, so we just had to roll with the punches. Off I went to Afghanistan along with my fellow teammates.

Waiting to leave felt like it was taking forever. In that moment you just want to get on the plane and head out. Sitting around and waiting just makes you think of all the time wasted in general. I wanted to get where I was going and lock my mentality in so I could focus on the missions ahead of me. I knew there would be time to experience other emotions later. I was in the process of flipping the switch.

The flight into Afghanistan was quiet. Not silent, just heavy. The kind of quiet that wraps around you when everyone's deep in their own thoughts, processing what it really means to step into war. No more drills. No more simulations. This was real now. The next time a shot was fired, it wouldn't be blanks. People would be trying to kill us, and we had to be better than them. Everyone always talks about going to war and what they will do and who they will be. Well, it was time to find out.

It is a clarifying moment in life for anyone who has been in the situation. I know who I am. I know when I am being threatened, shot at, or even blown up, I am a protector. I do not sit by and let it happen. Fear doesn't take over, action does. Time to move. War felt like something I was already familiar with. I don't seek war, I don't relish it, but I am capable of thriving in it. This isn't always a good quality to have, but in today's world, I'd rather be surrounded with men like me than the alternative. Peace comes with cost. Someone must be willing to pay the bill.

They sent us out in comfort, at first. A nice commercial plane to Germany, real seats, warm food, snacks like we were headed on vacation. But once we hit the next leg, it changed. They loaded us into the back of a C-17 military cargo plane like equipment. Tightly packed, buckled in with gear on our laps. No more pretense. Just soldiers and steel. The closer we got, the quieter it got and when we landed, it got real.

Bagram Airfield, thirty miles north of Kabul, was our first stop. The cold wind in the high altitude hit us in the face like a ton of bricks. Some people think Afghanistan is a massive desert where it never got cold. Instead, it is a place of extremes. In the mountains when it was cold, it was *extremely* cold. When it was hot, you were burning up. The air was thick with dust and tension. The ground felt foreign. The sky looked different. Everything in my body knew this wasn't normal. We had to start grabbing our bags that were shipped ahead of us so our gear would be there waiting for us. We loaded up on ammo and medical supplies.

I did not know it at the time, but this would be the nicest place I would see the whole time we were in that country. Overall, it didn't even feel like a war zone or third world country. They had buildings, fast food, internet, and an MWR, which stood for morale, welfare, and recreation center. I spent my time there lifting weights, taking classes on the local explosive techniques, and teaching classes on bomb disposal and clearance procedures.

Overall, it didn't seem like a conflict area or developing country.

From there we headed to the RC East region. Sharana was our next stop. That's where we took over some vehicles from the unit we were replacing. The vehicles were in rough shape, so we had to go through and outfit them. On the inside of the route clearance vehicles, there were Velcro pads that stick all over the walls in case you get hit by an IED, or an improvised explosive device. People don't think of this often, but when a bomb hits your vehicle, it is possible for shrapnel to shoot through the vehicle. This includes things already inside the vehicle along with its own bolts and screws. The last thing you want is to take a shortcut or miss something minor and someone winds up with a bolt snapping and getting thrown through someone's head. This is how much the little things matter. "Attention to detail" is the number one phrase heard in a route clearance Sapper unit. Bombs require attention. All of it. There is no room for anything else in your mind. Time and place matter. Flip the switch.

The elevation was brutal, over 6,000 feet higher than we were used to. And still, there was no time to acclimate. Some of our guys were already in rough shape—going through withdrawals from alcohol or pills. Imagine going to war with a man sitting next to you and he's going through withdrawal. That was such a weakness in my eyes. I hated that I had to look over fucks like this while I was preparing for war. Fucking ridiculous. I felt like I might be better off going in alone.

As a combat engineer, I had one of the most dangerous jobs: route clearance. We were the ones who went first. We found the bombs. We scanned the roads. We were the ones driving over IEDs (improvised explosive devices), so others didn't have to. There were no speeches. No glory. Just nerves of steel, sharp eyes, and absolute trust in the soldier next to you. The job gets infinitely harder when you can't trust the soldiers to your left and right to perform. One of my so-called leaders was an ex-Sapper instructor. The Army issues special fabric patches called tabs that are worn on the left shoulder of a uniform and announce rigorous training soldiers have successfully completed; for combat engineers, becoming a Sapper-tabbed leader is a primary goal. This specific non-commissioned officer (NCO) was on "*Surviving the Cut*", a military documentary that used to be on Netflix. He was one of the first soldiers I met with that background, and I was not impressed at all. He used his tab and reputation to hide the fact that he was a horrible combat soldier. One of the main things he taught me is that tabs and awards do not make a leader or a warrior. It is a person-to-person basis.

I was still setting my own path. I was one of two soldiers sent for specialized dismount and room-clearing tactics. I would have followed younger, less experienced me into war any day over this overinflated "leader." Combat has no favorites. Some people look good on paper. They have the credentials, but they do not have the Spirit. They are sheep in wolf's clothing. I did not listen to men like this. I told them to stay out of my way. I would make sure the job was done properly, and they would make it home alive. The two staff sergeants I said this to agreed and left me be. The thing about being in a warzone is…. Certain courtesies go out the window. Step the fuck up or get out of the way. Awards and accolades won't stop an enemy bullet from ending your life.

That lesson hit hard and stayed with me. Real leadership doesn't strut. It listens. It sweats with you. It apologizes when it's wrong and stands tall when it's right. I learned quickly to stop looking for guidance and instead find it within myself. Every time we rolled out, I knew—it could be my last. But I couldn't fixate on that. I was focused on being proficient, filling gaps wherever they appeared. I was a driver, a gunner, and I dismounted on foot. I saw combat in every position, and I engaged in the way I was supposed to every time. I taught classes on demolitions and IED search techniques to Green Berets and private contractors. Our explosive ordnance disposal (EOD) team and I created fake lanes using disarmed munitions to train soldiers. You can't afford to get rusty, not when it's your boots that will be first on the dirt.

Ironically, it wasn't the new guys who cracked mentally. It was often the senior leaders, the ones who had seen too much. I remember sitting them down, walking them through our prep, reminding them of their own training. These weren't weak men. They were just full. Full of memories. Full of ghosts. I learned to lead from the bottom, offering calm reassurance to the lower enlisted, and strength to the higher-ups who had carried too much for too long.

We'd scan trash piles. Look inside potholes. Watch for loose dirt or cell phones held at the wrong angle. Every detail mattered. Every instinct could be the line between survival and another folded flag on a casket. We lost several good men. Not all of them were my closest friends but when the net went black (meaning communications to home were shut down), you knew someone probably wasn't going home. Then, from intel received and shared in our channels, you were briefed on the soldiers who passed away or got injured. You didn't know all these soldiers, but you did know it could have been you or someone on your team. You knew that there was someone not going home and a family with a gaping hole in it. Their stories helped the other soldiers look out for similar situations that could cost them their lives. The insight provided by their losses was respected, and the life they lived was honored, but the weight was still heavy for all. But nothing compared to the loved ones waiting for that next phone call, Facebook message, or letter.

Then others were men who laughed with you, who passed around pictures of their kids, who made dumb jokes just to make the heat bearable. And then one day—they were just gone, and you still had a mission to do because to neglect your responsibilities and grieve meant more death. Even the kids in this region were a threat. I won't use names because I don't have permission to do so, but I remember one mission clearly. A sergeant first class was handing out water in a village. A kid, maybe seven or eight, walked up smiling and then stabbed him in the neck. That SFC died from his wounds that day. The kid, he vanished into a crowd like nothing happened. It was a gut punch. A reminder that evil doesn't always look like what you expect.

There were bombs inside dogs. Bombs on people. The intel in the region was vast but never seemed fully accurate. The Taliban would use anything, and anyone, to kill and control. The Afghan people? They weren't our enemy. But they were often caught in the middle, used as shields, props, or bait.

I lived by a simple rule: 1 + 1 = 2. One thing out of place? Maybe nothing. But two? That's a problem. Fresh dirt and a guy filming you from a roof? That's a setup. You learn to trust your instincts, or you die.

I clearly remember one specific time we were hit by an explosion. It wasn't our first combat encounter, but it felt like our first failure. We were clearing Route One, near Forward Operating Base Ghazni. My friend Bynoe drove a Husky, a large vehicle designed to detect mines and explosives while minimizing blast damage. Bynoe was good at his job and always spotted an issue if there was one, but there was no way he could detect this particular bomb, because it was in a water canal several feet under the road, and therefore undetectable.

When the IED went off, everything exploded. My ears rang. My vision blurred. Blood filled my mouth. I blacked out, came to, and blacked out again. I was driving one of the RG31 armored trucks and the explosion caused me to slam my face into the steering wheel. That was the first traumatic brain injury I can remember.

In this attack the worst of it was the Buffalo, the biggest and heaviest truck we had, as well as the strongest and most secure. Because the bomb was hidden and packed so deep in the ground, the concrete from the road was turned into a pushing charge. This threw the back end of the buffalo up in the air until it looked like it was standing on its nose. Then it came back down on its side abruptly. The truck was flipped back over by our wrecker (a big truck we use to tow other vehicles if they get stuck or break down or have blown up) and towed back to the base after we neutralized any threat. We made sure to secure the area by eliminating all enemy combatants and ensuring there were no longer any explosives remaining that could harm any of the villagers.

Some badly injured soldiers were medevacked from the scene after we secured the area. At the time, I just wanted to get back to the mission. I didn't think about concussions, or memory loss, or migraines that would last for years. I didn't tell anyone the full extent of my injuries. None of us did. To admit pain was to be pulled from the fight, and, worse, to be judged for it. You'd rather limp silently than speak honestly.

The worst attack came at the end of our rotation, summer 2013. We were preparing to leave FOB Ghazni. Everyone was burned out, numb, just trying to stay alive long enough to go home. Then the base got hit. A VBIED (vehicle-borne improvised explosive device) parked along-side the exterior wall. Around 1,200 pounds of homemade explosives detonated. I was lying on my cot in a GP Medium tent when it went off. The whole tent buckled. The pressure wave knocked the breath out of me. I couldn't hear. I couldn't think. I wasn't even dressed. I grabbed whatever gear I could, snatched my weapon, and ran to the perimeter. Instinct took over. Training locked in. Then we found out one of ours was missing.

I went out with two others to find him. We moved through smoke, rubble, and blood. There were bodies everywhere. Local contractors. Enemy fighters. Smoke. Screams. The sky filled with dust and gunfire. The base was locked down. All units were protecting their own sectors. Once the wall was breached, enemy fighters entered the FOB dressed as Afghan Army soldiers. We had to be very careful not to pull the trigger on the wrong person.

Later, a Polish sniper informed me that he had me in the crosshairs of his scope. He admitted almost pulling the trigger, until he seen my tattoos and realized I was friendly. "That was brave," he said. "What you did. That's what real soldiers do." I didn't feel brave. I felt like I was just doing what needed to be done. But hearing it from someone with the whole battlefield in his scope… That stuck with me. We held the line. We stopped the breach. But we still lost people. That part never gets easier. That was Afghanistan. It didn't care if you were tired. Didn't care that you were almost home. War just kept coming. An Army of Ghosts.

There weren't constant firefights and explosions, though. You still had to do everything to prepare for war and the battles to come. You still had to wash pots and pans, clean your weapon, work out to keep your body and mind sharp. The trucks and equipment needed fuel and overhauling.

Then you had your off-setting accidents. For example, we had a massive rubber bag that sat on the ground for fuel. There are more technical terms, but no one cares about that until it is an issue. Somehow, water got in our fuel. To get new fuel, we ended up having to burn all the contaminated fuel off. We were on a tiny, isolated combat outpost (COP) called Bande Sardeh, with no support, just us. We took control of our own walls and towers and helped train our Afghan Army counterparts. We had to ration our minimal supplies, restocking them was a logistical nightmare because the surrounding terrain was so mountainous. It was winter, and the ground was covered in snow and ice. Our one daily meal usually consisted of something like a spoonful of eggs and two pieces of turkey bacon or one small circle of some sort of sausage. As you can imagine, tempers were thinning each day when time finally came time for that one measly meal. To make things worse, our clean water supply was low.

The day the fuel needed to be burned was a really dreadful day. We had to monitor and control the burning of over 1,000 gallons of fuel. The fumes made people sick; there was fuel covering the ground and in our gear. I spent thirty-six hours straight working on this project to ensure the rest of the guys did not have to go through the same situation I was dealing with. I felt like I wanted them to focus on the mission. I knew I could manage the work and still be combat effective.

I wish I remembered it all more clearly. But over time, the memories fade. Or get scrambled. I ask old friends for details sometimes, but even they don't remember everything. Most people never heard my full story because I never really told it. I didn't want to. Not because I was hiding something, but because putting it into words makes it feel like I'm asking for pity. I also don't see it as my story to tell. Just something I lived through. My wife was pregnant while I was gone. Her family didn't understand. My own family didn't know how to process it either. I could feel the resentment building. like it was my fault for being away, like I chose war over them. I wasn't a different person. But war has a way of rearranging your pieces.

After that deployment, I slept less. I talked less. Trusted less. I wasn't angry all the time, but the anger was always close. I didn't want pity. I didn't want advice. I didn't want to explain how it felt to clean up what was left of a friend. Because honestly, how do you explain the sound of someone dying when it stops three feet from you?

You don't, you just keep moving. War made me sharp, efficient, and ruthless when necessary. But it also pushed me to the ease of violence. And that's the real danger—because you get so good at surviving, you forget how to live. Coming home was harder than combat. No one talks about that. Not honestly. Not without turning it into some twisted competition of "Who had it worse?" Even vets can be cruel to each other. Instead of support, it becomes sarcasm. Instead of unity, it becomes isolation. So, I stopped talking. Because most people's responses feel hollow. Empty. Like throwing a stone down a well and hearing no splash. But even in that fire there was growth. Brotherhood. Clarity. Purpose. I learned that courage isn't the absence of fear, it's the decision to move anyway. I learned that grief doesn't care about schedules. I learned how to lead in chaos, how to carry pain without dropping it on others. I learned lessons that would carry me in the life to come. But the cost was real.

My mission wasn't just combat, it was mobility, counter mobility, and survivability. Protection. It was service. To my brothers, to the local civilians, to a bigger belief that someone had to stand between life and the things that threaten it. War was simple. I knew my place. But the real battle, the one I never saw coming, was waiting for me when I got home.

CHAPTER 6

Coming Home–The War After the War

Only three people met me when I got off that plane coming home from Afghanistan: my mom, my stepdad Troy, and my friend Matt. No fanfare. No signs. Just open arms and presence. And I was deeply grateful because showing up for those you care about, matters. Especially when it would've been easier not to. Coming home should've felt like relief. Like closure. Like I'd crossed the finish line and could finally collapse into peace. But the truth is, it didn't feel like anything. Not at first. There's this idea that getting off the plane is the moment it all ends. You see it in movies, slow-motion hugs, kids running into your arms, someone crying happy tears. What they don't show you is what happens when the hugs are over. When the crowd disperses. When the noise dies and you're left staring at a life that doesn't quite fit anymore.

Coming home wasn't an ending. It was the beginning of a new kind of war, the kind you fight inside, in silence, while the rest of the world keeps moving like nothing happened. Physically, I was beat up. Neck pain. Back pain. Nerve damage made my left arm go numb. But I downplayed it. I told myself it could wait. That was my mindset: survive now, deal later. What no one tells you is that "later" always comes, and it rarely shows up gently.

The pain got worse. I started losing sensation in my hand. I couldn't grip or hold things like I used to. Lifting anything triggered sharp, radiating pain down my side. Simple things like holding a cup, tying my boots, tossing a ball with my kid, all turned into small battles I didn't talk about. Physical pain is easier to carry than putting it on others or being told you can no longer perform the job you've trained for. And emotionally? I didn't feel like I had space to speak.

While I was overseas, Caitlyn, my future wife, gave birth to our son Tristan. The pregnant woman I loved gave birth while I was in a Warzone. I missed that moment. That pain, hers and mine, never really seemed to heal. She didn't come to the welcome-home ceremony. I didn't get to hold my son for months. And when I finally did, it wasn't the warm reunion I'd imagined during those long nights in the desert. It felt cold, distant, heavy. She blamed me for being gone. For missing the birth. For not being able to fix it all when I came home. Over the years, that blame calcified into our rhythm. It became the air we breathed: strained, silent, sharp-edged. No matter what I said or how many times I tried to explain, it always came back to this: "It was your fault." As if I chose war. As if I wanted to miss everything. Her family wasn't very different. Neither was mine. I came home and walked into a silence that was louder than any firefight. They all expected me to be the same. Smile. Laugh. Let it go. But I wasn't the same. I couldn't be. My sisters didn't want to hear about Afghanistan, not even the funny stories. My mom cried when I told her things, even the light stuff. My wife changed the subject. Her grief was loud. Mine? Mine wasn't allowed.

It felt like I was being punished for surviving. People ask what changed in me. The truth… I don't really know. If you want to know what has changed, you'd have to ask the people close to me. All I know is I kept growing and kept evolving… but they wanted the old version of me back. That version didn't exist anymore. You can't unsee what you've seen. You don't come back from war and press "reset." You either move forward, or you break. Sometimes, moving forward *is* letting old parts of you die so you can become something more equipped to live this life. Quietly. Piece by piece. Internally. Not with a bang, but a slow leak of self, until you don't recognize the man brushing his teeth in the mirror. When I came home, my body was failing me, my family didn't understand me, and the world felt too fast and too shallow. But I still had to show up. I had kids who needed me. A future that still had to be built. A life that meant more than just surviving Afghanistan.

Wartime Jacob was a completely different person from peacetime Jacob. I couldn't be wartime Jacob, at home. People who have experienced combat must have different reservoirs for different severities. The mindset needed to survive combat is not very useful at the grocery store. Did I have moments when the two worlds slammed into each other? Of course I did. Moments like being at the grocery store, realizing I didn't have my weapon, and then racing back to my car to grab it. I lived with my weapon on my arm and in my hand for so long. It was part of me, like an appendage. Once I realized I wasn't carrying it, it was like an alarm went off in my head. It's funny to think about it now but then it felt embarrassing. I would look around to make sure no one saw me frantically running back to the car, try to laugh it off, and just keep living.

Driving could be very alarming as well. Hitting a speed bump or a pothole in the road quickly felt like the end of the world, far too similar to the initial jolt of a bomb going off under your feet, and it could ruin your whole fucking day. Not just from your own reaction, but from the reaction of the people around you. You feel your body shift in response to the shock, so you brace for impact, hold your breath, grip the steering wheel like you want to choke it to death, and maybe even swerve. Now your heart is pounding, the ringing in your ears keeps getting louder, you're breathing like you just went ten rounds with Mike Tyson and that's when it all gets worse. Say you are in the car with your family. Everyone panics in response to your response. Confused and upset, they scream in fear or start yelling at you about your shitty driving. This does not help you calm down. If anything, it amps you up more. They do not understand that the war version of you is screaming at the thin glass that separates the docile version of you from the "violence of action" version of you. Now, in a split second, you must subdue the side that has kept you alive and simultaneously calm everyone else down. You must make them feel better about your shit. You don't have time to put the scary parts of you away before you start telling everyone to shut the fuck up and be still.

Sleeping proved to be even more challenging. I was used to the sound of mortar rounds flying over me and crashing in the distance. The sound of generators running all night to keep operations alive. The sound of "INCOMING" playing over the makeshift speaker system through the Forward Operating Base (FOB) or Command Outpost (COP). The silence seemed too loud and made it hard to fall asleep. If I managed to fall asleep, I had trouble staying asleep. Noises like babies crying in the night could really wake you up in a weird head space. Feeling someone moving around in the bed you share did the same thing. You have no control over the sleeping you. I could maintain composure when I was awake. But when I was asleep.... That was a different story. It is a wild feeling to fear yourself. I would not let my children sleep in my bed. Especially not the baby. The risk was too high. Yes, you can flip the switch but that does not mean you are always fully in control. The moment your mentality switches from peace to action, the body and reflex take control. Like getting startled and flinching. Except with this flinch, you have an unwavering need to react and move to assault the threat. Even if there isn't one. Now you need to turn yourself inside out and put all the action, aggression, and rage back to the other side of your being so you can bring back the reasonable person that was just startled for a bit. Like caging a great white shark after it tasted the first drop of blood.

There were nights I'd sit in the living room with the TV on mute just to fill the room with light. I couldn't sleep. I couldn't stop scanning. The sound of a trash truck at 3 a.m. would jack my heart rate up like a rocket. It wasn't fear. It was conditioning. Reflex. Muscle memory burned into my bones. Some mornings I'd drive in circles before work because I didn't want to sit at home while I could not sleep. I couldn't get sleep there and the Army required me to show up for physical training by 6AM anyway. Home didn't feel safe. Not because of violence—but because of me. The emptiness of not being the husband I was. The father they expected. The man I used to be. People think combat veterans are angry all the time. But anger is just grief wearing armor. I was grieving things I couldn't name—lost time, fractured relationships, parts of myself I left behind in the dust.

I tried therapy. The counselor did my intake and scheduled me for a different appointment. The second appointment, the counselor just gave me pages of paper to read and asked whether I had experienced "trauma." I refused to use that word because I didn't feel traumatized. Trauma is a watered-down word in this world. It is an opinion and people adapt it to what they want it to mean. I personally dislike the word, and it makes it difficult for me to stay on track when it is said to me. So, When I said no, we stared at each other in silence for thirty minutes. I didn't go back. Not because I didn't want help—but because I didn't know *how* to be helped. I had no respect for the people in the sorts of positions that were supposed to help. The therapists were there, but they did not provide help, nor did it seem like they even wanted to.

Maybe it was just me. I just didn't relate to that sort of help. I've spent years understanding myself and my emotions. The truth is this: those who want to understand don't point fingers. The people whose job it is to help are not just there to help. They are analyzing you to send a report about you after. They seem more interested in the wrong things like how to point the blame at something other than the military. They were trying to make connections with what could be blamed on you rather than the military. At the end of the day, they worked for the Department of Defense, and their job was to protect the department more than help the client. Health care doesn't work for the patient that way. I made excuses for people I cared about. For why they refused to try to understand or be there with me. Everyone was more interested in how I affected them, while no one really cared about how the interactions affected me. I spent my whole life swallowing myself so others wouldn't feel bad about themselves. I tried to sleep correctly for years. But sleep didn't come. Not real sleep atleast. Just these shallow, naps full of fragments and static.

What saved me wasn't a breakthrough or a program. It wasn't a moment of divine clarity. It was me choosing to make every day a new day. I had shit to do regardless of my feelings or thoughts. Your thoughts are not you. Thoughts are tricky. You must learn how to shift gears. Say no. Do it anyway. My thoughts are not real externally, only within myself. Thoughts and feelings are internal and intrinsic. A feeling doesn't make reality and the truth is, I control my actions. I control my mind; it does not control me.

It was small, stubborn acts of rebuilding. Taking my kids to school. Saying I love you, even going for walks. Going for runs when I didn't want to be alone. Being honest, first with myself, then with others. It was messy. Uneven. Slow. I even started writing again. Not for anyone else. Just for me. Scribbled thoughts on napkins, notebooks, margins of old receipts. I didn't write about war; I wrote about now. About trying to fix a broken sink without cursing. About the way my son giggled when he watched *The Lorax*. About sitting in a grocery store parking lot for twenty minutes because I didn't want to go inside. It helped. Not right away. But eventually. Eventually, I stopped feeling like a ghost.

I stopped waiting for someone to meet me halfway. I realized no one was coming to fix me, and maybe that was OK. I didn't want anyone to. I am not broken, maybe that was the point. I like the way I am, and I've accepted how I want to be. Yes, I can be a lot to take in. I am an aggressive person, I do not like to accept excuses, and I try to push those around me to help them see themselves truthfully, but I have practiced speaking for many years now and the messages are clearer, the delivery is more refined I am aware I can be overwhelming for some.

Because coming home isn't about returning to what was. It's about building what *is*.

Even if you have to do it from the ground up. Even if your hands shake while you do it. Even if you're scared most days and don't recognize your own voice.

I'm still rebuilding. Some days, the structure leans. Some days it collapses. But every day, I try. And that has to count for something.

I didn't come back just to disappear. I came back to live. And maybe that's the bravest thing I've ever done. Death is easy, living is hard. There's a unique kind of war that starts when the battlefield ends. It doesn't come with gunfire or commands shouted over the roar of helicopters. It starts quietly, sometimes with a ringing in your ears that never goes away, sometimes with a headache that builds until it feels like your skull might crack from the inside. Sometimes it starts with nothing more than the stubborn need to keep moving forward, even when your body is no longer built for forward motion.

Traumatic brain injury (TBI) is a clinical term. I didn't name the injury but it doesn't convey what it feels like to wake up every morning with a migraine so sharp it makes you nauseous. It doesn't explain the sudden waves of aggression, the moments where patience collapses into fury without warning. It doesn't account for the panic in your partner's eyes when you're not acting like yourself, and the shame you feel when you recognize it too late because the damage has already been done. The damage in the perception of how your partner views you.

The explosion that knocked me off my axis left more than just ghosts in my head. The shockwave tore through my spine like a blade. Hemorrhaging discs in my neck and lower back, pinching nerves and slicing through sensation like threads in a web. The left side of my body hasn't felt quite right since. Dull, slow, and unpredictable. My hands are the worst. There's a disconnect between what I want them to do and what they can manage. Turning a doorknob or holding a cup without dropping it is a minor miracle some days. We all have good and bad days so we must deal with them head on. In my youth, I did not fully understand the impact of combat and the toll that being in an explosion took on me. Now I have a clearer image, and I can't say I would have chosen differently. I'm at peace with my choices.

Every morning, I still lace up my shoes. I still go for the run. I still hit the gym, not because it's easy, but because it's necessary. The body might scream, and the nerves might misfire, but motion is survival. If I stop moving, I stop being a father, a husband, a man who contributes. I'll never be the guy I was before, but I can be something else. Something steady and persistent. Something that doesn't quit just because it hurts. And it hurts. Every day. But that pain is irrelevant. Life is painful. There is no way around it. The pain is worth the life I am living. I'm grateful for every single day that I get the chance to live. I'm living on borrowed time as it is. So, gratitude is all I have to give. I made a choice. I won't mask the pain with pills like the medical system insisted on. When I first reported the headaches and back issues, they offered opioids like candy. No real plan. No long-term rehab. Just more pills to quiet the symptoms. I've seen too many brothers disappear into that spiral, medicated into numbness until they don't know who they are anymore. Until the pain becomes just one more thing buried under dependency.

I chose the pain. I chose clarity. But that choice isn't heroic, it's exhausting. It means gritting my teeth through family dinners. It means nodding through bedtime stories with a migraine thumping behind my eyes. It means hiding the twitch in my left hand when I help my kids tie their shoes. It means pretending that none of it is happening, so everyone around can live the same life that they were living. It means these are my issues to bear so I need to ensure I don't accidentally push my strife onto others. I chose my life, and I chose my path. Now it's my responsibility to live with. I carry my burdens willingly and gratefully. And still, every morning, I hit the day with a solid workout. Because I can. Because I must.

A certainty that I've come to realize in my life is that the hard path to take is always the right path to choose. It's harder to read a book than to watch a movie, so reading is probably better for you. Hard is good. Hard leads to growth. Growth starts at failure. I believe a person should accomplish one hard task per day. Just to keep their mind resilient and their body strong. I hate reading so I force myself to read. I hate running so I force myself to run. I've lived my life at a certain velocity and now I can never stop. Because the moment I stop is the moment everything will give way. It's not motivation; it is dedication. The feeling of motivation is wanting to do something. When people rely on motivation, it fails them every time. I don't know what day, what hour, or what minute, but it will fail.

Do what you need to do because you should. Your feelings are not valid, and they do not trump the fact that shit still needs to get done. No matter how I feel about the work, it won't change the work that needs to be done. Our blessings from God wait in the hard work before us, and we will never get those blessings if we refuse to do the work that unlocks them.

CHAPTER 7

Leadership Expansion

I used to think that leadership was about being aggressive. That command came with a voice raised above the chaos, that aggression meant authority, and anger meant strength. Early on, I let my temper do the talking. When things didn't go right, I let it rip. When lives were on the line, I just could

not understand why people didn't take the training or the learning or the fitness readiness seriously.

One day, after a particularly brutal training exercise, I caught my reflection in the mirror of a dusty Humvee. My face was red, my voice hoarse, and my soldiers had that familiar look in their eyes, not of respect, but discomfort and fear. They weren't leaning in to listen. They were trying to just stay out of my way. I didn't look like a leader. I looked like a man who'd lost control. This was very early in my career, and I was extremely angry. During a dry run of a breaching mission, I had a soldier under my responsibility set his weapon in an angle that was pointing to our friendly support by fire element. We were about to do this in real life. Real bullets. I could not have such a huge mistake risk the lives of the rest of the soldiers. In my head, I didn't have time for words. We just blew a massive hole in the enemy compound wall and here we are with my soldier standing in the breach with his weapon pointed right at our own people. This was unacceptable so I kicked the weapon down, grabbed the soldier by his chest and arm, and slammed him to the ground while screaming at him.

This was not the right reaction. But at the time, all I could think about was having to carry the weight of one of my soldiers accidentally killing one of our own. In my eyes, my reaction ensured he would never forget and was better than him living the rest of his life in regret, along with one of our own not going home. Most of my leaders were lazy and out of shape. They were not willing to learn new or different concepts from the "old ways", and just wanted to push for retirement. I was training for war, and they could care less.

It was a hard lesson, but one I needed: yelling and being aggressive when you're angry doesn't command respect. It just makes you look silly. Worse, it drains your credibility. Leadership, I came to learn, it's all about the weight of your presence. So, I flipped the script. I stopped yelling when I was angry. Instead, I learned to pretend to be angry when I was calm. Controlled, deliberate anger—that's different. That's a tool. When used sparingly, it cuts through the noise like a knife. When I was calm, clear-headed, and knew exactly what I wanted to say, I let my tone carry the message. Not my emotions. Not my ego.

I also learned that fear—the kind leaders rely on—has a very short shelf life.

Because once you've had bullets flying over your head and seen men duck for cover as rounds crack past their ears, you realize something: nothing I could ever say, no punishment I could ever dish out, would ever scare my soldiers more than the battlefield itself. Real fear lives out there, beyond the wire.

So, if I wanted them to follow me, it couldn't be because they were afraid of me. It had to be because they trusted me. Trusting that I had their best interest at heart. That every decision I made was made with their survival, their success, and their growth in mind. That trust, not fear, is how wars are won.

And here's another truth I had to wrestle with: no two soldiers are the same. You can't lead every person the same way and expect them to thrive. Some need a firm hand. Others need patience. Some want to be challenged, some need encouragement. A real leader learns the language of each member of the team. You must tune your leadership style like an instrument, adjusting it with each personality, each strength, each weakness.

That's when I started to understand what leadership really was. It's not about being the best. Hell... I'm not the best at most of what I do. But I am the best at seeing what others bring to the table and putting them in a position to succeed. I can build a team like nobody else. And that's what matters. Not being the strongest, or the fastest, or the smartest, but recognizing talent, nurturing it, and connecting the right people with the right roles at the right time. Leaders must also exhibit true willingness to do and live everything they would ask someone else to do. For example, my son Tristan runs cross country. On the days I make him go for a run, I run after him and twice as far. I don't run with him because I do not have the goal of being a competitive distance runner. I know I would just slow him down so to make up for the speed, I go farther. I explained to him that I'm not just telling him how to push and succeed, I'm showing him. I'm not telling him who to be, I'm just showing him who I am and giving him the space to follow. A leader does not choose their mentees; the mentee must choose the leader.

Leadership isn't about you. It never was. It's about what you can unlock in others. It's about building something that lasts, something that functions even when you're not in the room. That's the mark of real leadership—not the footprint you leave behind, but the steps you help others take forward. The highest priority of a leader should be to make themselves obsolete. Not out of negligence, but out of training, coaching, and mentoring in a way that leaves no gaps. Lead so well that if you are gone tomorrow, the people you led can pick up right where you left off and nothing fails in your absence.

I always told my soldiers, "I will be the first man in the door every time. You are all trained well enough to do my job and I cannot carry all your bodies home if something goes wrong, but all of you can carry me home if need be." My life as a leader was not more important than the soldiers under me. If I did my job right, my life was less important because leadership is responsibility. Responsibility is a privilege and a curse.

So now when I lead, I do it quietly, but intentionally. I listen more than I speak. I watch more than I react. I know when to push, when to pull, and when to just stand beside someone and let them figure it out. Because leadership isn't about being in front. It's about bringing others up to the front with you. And when I look back at my proudest moments, they're not the medals or the missions. They're the people. The soldiers who went on to lead others. The ones who learned to believe in themselves because I believed in them first. That is what matters. That is how battles are won, and more importantly, how leaders are made.

Leadership has many faces. It wears crowns and combat boots, speaks softly or roars, watches in silence or weaves webs behind the scenes. But at its core, leadership always reveals the character of the person carrying it. So, ask yourself, what are you? The lion, the elephant, or the fox? The lion doesn't speak first. It doesn't need to. The lion's leadership is rooted in its presence, undeniable, magnetic, commanding. People respond to it instinctively, even if they don't fully understand why. You know when you're in the room with a lion. You feel it. This leader isn't constantly reminding you of their position or title. They lead by simply being, by standing firm in who they are. Their authority is rarely questioned, not because they are feared but because they are clear, clear in purpose, clear in direction, and clear in character. But don't mistake that silence for softness. The lion's wrath, when provoked, is real and swift. They don't fight to prove a point. They fight to protect what matters. Lions lead through presence, not persuasion, through conviction, not coercion.

Now the fox—ah, the fox is clever. Witty. Strategic. But the fox views leadership through a different lens: control. For the fox, it's not about vision, it's about leverage. It's about finding the psychological cracks in others and slipping in quietly to gain advantage. The fox doesn't lead people. It steers them. Coaxes them. Manipulates them, sometimes with charm, sometimes with guilt, often with half-truths and selective transparency. And here's the tricky part—the fox often thinks this is what leadership is. Getting people to do what you want is the goal. If they don't resist, you've led well. But that's not leadership. That's puppeteering. I call this the fatal funnel of leadership: when influence becomes manipulation and people become pawns. It worked for a while. But it always crumbles. People eventually see the strings. They always do. And when they do, the trust evaporates, and the fox is left with no one to lead.

This type is the most common today. In the military. In business. In everyday life. It's everywhere. And the most dangerous part? The fox doesn't see its own weakness. It believes its tactics make it strong.

But true leadership doesn't come from tricking people into following. It comes from caring enough to understand what drives them—and investing in that. When people know they matter to you, they will choose to follow. Not out of fear. Not out of reward. But out of belief.

Then... there's the elephant. The elephant doesn't push. It doesn't shout. It doesn't have to win anything. The elephant just is. It doesn't need to compete because its power is in its pace. The elephant moves with quiet certainty, and when it moves, entire ecosystems adjust. It is not reactive; it is steady, thoughtful, and deeply aware. This kind of leader doesn't force their way into the spotlight. They grow into their space, and that space naturally expands. The elephant doesn't need to remind you how big it is. You just know. And yet, with all that mass and might, the elephant is graceful. It knows the weight of its steps. It walks carefully, not because it's afraid, but because it respects the ground it walks on—and the people walking beside it. The elephant leads not by command or control, but by example. By patience. By wisdom. It is deeply rooted but always evolving. This leader has the rare gift of moving forward while remaining grounded.

So again, ask yourself: Do you command attention like the lion, without needing to roar? Do you manipulate, like the fox, mistaking control for connection? Or do you move through the world like the elephant—aware, intentional, unshaken? Every leader leans toward one of these instincts. And most of us cycle through all three at times. The key is awareness. But if I had to choose...

I'd follow the elephant.

I'd listen to the lion.

And I'd never trust the fox.

Leadership isn't about being the smartest or loudest in the room. It's about being the kind of presence others grow stronger around. So again, I ask you: Who are you?

CHAPTER 8

Starting Over—Real Estate, Mistakes, Lessons, Growth

After everything I'd seen, everything I'd carried, reenlisting might sound strange—but for me, it made sense. From 2012 to 2015 I was stationed at Fort Drum, New York. I signed another contract with the Army at Fort Drum to be transferred to Fort Stewart, Georgia. I didn't have anything left for New York. That place drained me more than Afghanistan ever did. The cold. The isolation. The feeling of being stuck in a place that didn't want me.

I was a sergeant now. I'd earned some stripes, some authority. And with that came the chance to build something new. I wasn't running from war, I was running from the silence of home. Because at least in combat, I knew who I was.

Home hated me. War welcomed me. In Georgia, something shifted. I bought my first house in 2015. Sight unseen. To most people, that sounds reckless. To me, it was bold. Necessary. Liberating.

That house wasn't just shelter; it was a declaration. The first thing that was *mine*.

It didn't matter that I had not walked through it or inspected a single corner. What mattered was that I finally had a place where I could breathe. A place where I didn't have to look over my shoulder. For a kid who grew up listening for footsteps and slammed doors in the middle of the night, that house was more than a roof. It was safety. It was *freedom*. It was a start.

That house became the foundation for the man I was trying to be. Not just a soldier. Not just a survivor. But a father. A provider. Someone who showed up, not perfectly, but persistently. Someone who had pain in his spine, war in his memory, and a fire in his chest—but kept moving forward anyway.

My kids became my why. Fatherhood became the new battlefield—and I was determined to win. Not with authority or fear. But with *presence*. With patience. With consistency. I didn't want to repeat the story I came from. I wanted to write a new one. One where my kids saw a man who had walked through life and chose to live. A man who could be tough without being cold. Beaten up, maybe—but never bitter. I wanted them to see that healing is possible. That growth isn't weakness. That you can carry pain without letting it define you. Because here's the truth: *I love life. I love every single day of it.* Not because it's easy. But because I know what it's like to live without and I have watched young men die before it was their time. If I hadn't gone hungry overseas, I wouldn't appreciate a stocked fridge. If I hadn't slept in dust and heat, I wouldn't have treasured the feeling of my own couch. There's beauty in the ordinary, but you only see it if you've suffered long enough to understand what a gift it is.

Struggle sharpens the lens. Pain refines your vision.

Most people live life chasing the next thrill, the next fix, the next big thing. But when you've spent years in grief, in silence, in survival, just having a safe place to lay your head becomes holy. I started to understand: Value doesn't come from a rank, medals, money, or status. It comes from presence. From *character*. From showing up when no one's watching. From doing the right thing, quietly. Buying that house was the beginning of something. The first time I realized, I didn't need war to have purpose. I didn't need someone else's permission to start over. I just needed the courage to build.

CHAPTER 9

Engineer Explosive Ordnance Clearance (E-EOCA)

At this time in my career, a course called "engineer explosive ordnance clearance agent," was one of the most important classes I could take. This gave me the ability to be the person on the ground to decide what was going to happen next. In route clearance, this meant not having to wait for hours on end while Explosive Ordnance Disposal (EOD) teams to come see what we are dealing with. Before we move on, let me explain a few things. Route clearance is a wild job to perform in a warzone. The job requires a Combat Engineers to travel down roads, or paths that are suspected to have improvised explosive devices (IED's,) in the ground. They can be activated by a timer, radio wave or signal like a cell phone, battery pack, pressure plate, and a few other ways. The goal is to find the bombs and either dispose of them or render them safe. In route clearance you drive along the route going about two miles per hour in hopes that you see the threat before it blows you sky high. We drive and walk these roads to clear them of the hazards, so others do not have to. This is arguably one of the most dangerous jobs. As you can imagine, when you do find a live IED, you want to move quickly and precisely to avoid getting ambushed or attacked. The last place you want to be in when you are getting shot at, is handling a bomb that is designed specifically to kill you. Time on target in a combat zone is essential. This means once you find your target, you start watching the clock. There is no time to waste. The problem with this along with bomb disposal is that you do not want to rush it when any second could be your last. You want a calm mind and a steady hand. Anyone can pull a trigger when they are scared or defending themselves. Handling a bomb with precision and a calm mind? Not so much. This isn't some *Hurt Locker* bullshit. Bomb disposal is not as easy as yanking on a device and pulling several armed missiles out of the ground and just cutting wires. In the real world, the shit you see in movies would have gotten everyone killed. Some have anti tilt or antihandling devices. If the munition is moved at all... It will explode. Others have a battery pack or capacitor to store power. If you cut the wire, the energy is dumped into the IED and again, everyone dies. Bomb handling is an art. And this course would make me an elite artist of my craft. For a Combat Engineer, the course would make me more efficient in my overall profession. In Afghanistan, whenever we would find a suspected IED and it was out of our capabilities to dispose of, we would have to call in the EOD team (our version of a bomb squad) and it could take hours for them to arrive. We would have to secure the area and wait. This is just not an ideal way to handle a situation where every second counts. I was already trained in combat and handling IED's, but this was just a higher level of knowledge and experience that would make me better at ensuring the safety of troops along with the local populous.

We learned about types of munitions and triggers, the inner workings of it all, from all over the world. We even learned about guided seeking missiles ... But that is something we will not discuss here. We learned how to handle these devices, which ones can be blown in place (BIP) or the ones we would need further assistance with from EOD. We learned about the bomb investigating robots including the Talion and two other lighter robots that could easily be carried by hand. The Talion was by far the best approach. This was a remote-control robot that sat on a track. It is pretty large but not so big that it could not be picked up and moved. If you like watching action movies, I am sure you have seen them on some show while it was being used by SWAT. In my eyes, this course was the difference of life and death in most scenarios for a Combat Engineer. The curriculum was ordnance identification, safety precautions, reconnaissance and reporting of ordnance, and protective works and demolition techniques.

Identifying ordnance by classification relies on several criteria. Shapes and sizes reveal a lot. Understanding the color codes for different countries and the universal ones lets you know the intent and impact of the munition. All bombs found in warzones are not IED's. When you think of a country like Afghanistan, you must remember their long history of war. That country has been in several different conflicts so there are bombs there that have been there for many years before we even got there. Sometimes a rocket, missile, or land minefield would sit and just never explode so no one ever messed with it. Over time these things can degrade and explode at any minute. Keeping that in mind, another job as a Combat Engineer is to clear these bombs along with mine fields. So, laying on your stomach, digging up land minds and de fusing them, moving them in a way they don't explode, or clearing the area and detonating everything to make sure it will not randomly explode and kill people by accident. This means other Countries will label and color code their munitions, so their soldiers know what they are firing off. All countries have a marking and labeling system, so it is important to learn the systems because all explosives are not handled in the same manner. A huge part of this is understanding the fuse and initiation system. Mostly, if you can remove the fuse or stop it from triggering… you can stop the threat. Fuses go into several categories. Some detonated by burning through flammable powder, by a spark or a tilt. Some know its surroundings due to sensors in the tip that can read the brightness of the sun. Kind of like cars when the lights automatically turn on at night. These are called proximity fuses. They usually had some sort of clear or light-colored cap, and they could see shadows. It registered the sunlight and could adjust on sunrise and sunset but any quick change in lighting or surroundings and it was game over. With these you had to understand your positioning along with the angle of the sun. you wanted your shadow casting away from the area and not into your working space.

Several were set for time. The fuse head rotates for a delay and are set for time frames to let the munition hit its target and blow once it could have the most killing and damaging effect. They were not set to detonate just on impact alone. These fuses were metal. You could see the time markers on the sides and in my opinion were the easiest ones to disarm. The only issue with that is, you have no idea how much time you have until it goes off. There is no digital time screen counting down seconds. The only way to see it is to get your face within a foot of the fuse and watch the dial twist and hope you have more than ten seconds left.

Can you imagine this? Put yourself in this situation. There is a bomb lying on the ground and you must trust yourself and your knowledge enough to approach it and handle it. You have no idea when it is set to go off or if it will ever go off. You have to approach it in order to find out how to handle the situation. You can understand the mental and physical conditioning you need to do this without letting your thoughts run. No time for "what if's." No time for goodbyes. No time. There is just you and this moment. If you want to come out alive and without getting anyone else killed, you have to let your life go. Let go of everything you care about in order to preserve it. There is a place in your mind you leave yourself. Put it away, leave it at the door. Only hold in mind what is needed in that moment. When it's over, you can go back to being a human. Flip the switch. This course had a huge impact on me and my mentality of the tasks at hand and the missions to follow. With knowledge I felt armed. I was confident in not only my ability for physical action, but in mental conquering and precision in thought. There are different versions of me, and I am in control of when I let them out. This course made me more effective in combat and better suited to protect the Soldiers to my left and right. All this information would be used in my military career to teach and mentor my junior leaders as well as be combat effective in the ongoing war in Afghanistan.

Protective works and demolitions were probably my favorite part of this course. Most people didn't like it because it was quite a bit of manual labor, but I loved it. This is where you would improvise. Say there is a small UXO (unknown explosive ordnance) close to a farm or a wall to a school. First you would identify the object. Then you would decide if you were going to blow in place or not. If we decided we could, there needed to be protections made. One way to do this would be to fill sandbags and place them around the wall until the trajectory of the explosion would not affect or hurt the structure. This was a great way to show the local populus that we wanted to keep their community safe. During the course, we were able to set this up in real life. We went to the demo range on site and practiced different formations and shapes to blow shit up and see what worked best. We had several munitions and a bunch of c-4, Det cord, Time fuse, and shock tube. This made for a fun day of walking, building, and demolishing. C-4 is the military's primary source of explosives for several reasons. It requires a good amount of heat and pressure to explode. This is optimal for soldiers to carry because you would not want to carry and unstable substance in a bag as you climb mountains and get shot at. C-4 can be grabbed, squeezed, thrown, or stomped on and that will not make it explode. You can even use a lighter and set it on fire and it still won't explode. Det Cord is what we use the prime the C-4. The det cord packs a big enough explosion to get the C-4 to explode. Det cord is similar to C-4 because it takes quite a bit to set it off. It is a green tube or rope-like sheath with explosive grains inside of it. Now the Time fuse is different. It is shock tube with a blasting cap on the end. The shock tube can be lit, and it burns through instead of exploding like the other 2 elements I have just described. All these items together and you can set up the charges you need to while ensuring safety and distance in a controlled environment. There is much more to all this, but I want to respect the lessons I have been taught and there is no real need to go into further detail. Any more detail than this and it would be in the risk of operational security as well as sharing classified information. This is not a book to share military secrets. It is a book of growth and respect for life.

CHAPTER 10

A New Mission—Becoming a Career Counselor

After coming home, after the bruises, the numbness, and the resentment settled into something like silence, I realized that changing where I lived wasn't enough. It was 2017 now and I needed to change how I served. I had been a combat engineer, an instrument of war. Built to lead from the front of a mission, clear routes, disarm bombs, carry the weight, take the hits, and keep going. That role forged me. It gave me purpose when I needed one. But I knew I couldn't carry that version of myself forever.

I didn't just want to survive… I wanted to evolve. I found myself missing the warzone. My life at home felt fake. It was hard to care about daily things. I knew I had to separate myself from the fight so I could stay in it. I knew something had to adjust in me. A life waiting for the next war is a life far from what my children needed from me. That meant serving differently. With my mind, not just my body. With perspective, not just power.

So, I made a move that surprised even myself: I reclassed and became a *career counselor*. The process of becoming a career counselor was a little unorthodox in my situation. I had the privilege of working with Master Sergeant Celeste Spencer. She was extremely knowledgeable and intelligent, kind and fair. She was strong in character and fun to work with. I didn't know it at the time, but she was going to help me change my future in the military, far more than I expected. Master Sergent Spencer was another person in my life who appreciated things beyond my physical abilities. She was more interested in my mental and psychological characteristics. She valued my words and input, and she saw where I could fit in being a counselor and what value I could add. Her recognition of my potential was a huge turning point in my life.

At first, my original packet was denied because the Army had enough counselors. When Master Sergeant Spencer told me I wasn't going to be accepted, I just thanked her for her time in allowing me to grow and teaching me new aspects of serving, in a different way. Later that same day, she called me into her office and let me know that she got personally involved to make sure my packet was pushed through and accepted. She explained that my calm acceptance in handling rejection while still being grateful for the experience of working in a counselor capacity was all she needed to see to know that she wanted me on her team. It was because of her alone, that this was possible in my life. She was not the first person to tell me I had worth outside of war, but her example was the first time I noticed a pattern in intelligent minds: the kind of character they wanted to surround themselves with. And that person was me. Not because I was special, but because I was grateful. Gratitude in life takes practice.

To outsiders, it might've looked like a soft landing. A desk job. A way out of the fight. But I saw it as a new front line, a place where the battles were quieter but just as important. Because soldiers aren't just shaped in combat, they're shaped in decisions, in doubt, in the small moments no one sees. This was my chance to help them before they hit their personal war zones. I didn't become a counselor because I was done serving. I became one because I wasn't done *caring*. And when I stepped into that new role, everything shifted. I wasn't guiding convoys anymore; I was guiding lives.

Helping young soldiers understand their benefits. Chart their career paths. Avoid the financial traps, toxic leadership, and silent spirals that so many of us fall into. I became the voice I wished I had when I was a nineteen-year-old kid about to enlist without knowing what all was offered, without care or direction.

For the first time, I saw that my experience wasn't just pain, it was preparation. I wasn't counseling with checklists or pamphlets. I was listening. Asking real questions. Helping people realize they had options, potential, and power. I wasn't just handing out advice, I was giving people permission to be honest with me but more-so with themselves. And I became damn good at it. Not because I had all the answers, but because I took the time to help others ask better questions. Soldiers started seeking me out, not for rank, but for real connection. My office became a place to breathe. A place where they weren't just numbers. They were people. I learned something in that space I'd never learned in combat:

I didn't need war to be valuable.

I didn't need pain to be worthy.

I didn't need to suffer to serve.

Becoming a career counselor didn't just change my MOS. It changed my role in leadership. It reminded me I was more than just a survivor. I was a teacher. A guide. A builder of people. I wasn't just carrying ammo and demolitions anymore. I was helping soldiers carry vision. In 2018, I transferred to Fort Campbell, Kentucky. The new team I was joining was a few people short of a full counselor team. So, I took the time and built my team. Luckly, I had a friend that was already at Campbell. He was someone I knew I could trust. His name was Darrell Vassey.

I met Vassey during the Combat Engineer Advanced Leadership Course. One day we were going for a run with the instructors before class. Most of the other students were trying to find ways to get out of having to go or an easier route. This frustrated me; I viewed this as a weak mentality and weak character basis. Vassey had a medical issue that prevented him from having to run that day, but he said he wasn't a bitch and was going to tough it out in order to stay with his classmates. I respected that decision more than he knew. Vassey kept up with us the whole run, tears in his eyes from enduring pain from his injured foot. At that moment, I decided we would become close. Vassey was a person I could build a strong team with.

In Fort Campbell, I brought Vassey on, and he quickly became one of my strongest assets. We secured thousands of contracts for the Army and helped so many soldiers either find a new job to stay in the Army, or continue to serve in the combat fields they were in. Our success came from the ability to relate with that group of light infantry soldiers and to understand their struggles.

Six months into being a career counselor, I had the honor of competing for the title of Career Counselor of the Year (CCOY) in 2018. Vassey was also chosen to represent us in the Retention NCO of the year competition. The competition closed with an awards ceremony. As we stood beside our competitors, in front of a room full of leaders awaiting announcements, Vassey and I were both able to take home awards for the Rakkasans. (An article about our win can be found by googling Rakkasans and our names.) I was proud to have coached and mentored a great friend of mine, a peer, and a fellow leader. I was not just validated in myself; I was validated in my ability to lead. Vassey winning his category meant so much more to me. It meant I did right by him.

Not only that, but I had also chosen the right man to dine at my table.

The best Sapper Competition was the only other competition that held weight in my eyes. It was my pride and joy. I was already set to change my job, which meant I would no longer be a light combat,

Sapper. Before I left, I signed up to compete to show my soldiers that pushing yourself never stops and to set the example that they should strive to learn and grow no matter what. When I was announced as the winner of that competition along with my partner, I was ecstatic. It was a seventy-two-hour competition that included rucking over fifty miles in entirety, doing several combat and demolition simulated tasks, completing an air assault mission, and doing a stress shoot with an m-4 and CREW weapons (automatic machine guns M249 and M240). It felt like a perfect way to leave combat and focus more on the readiness of soldiers and the Army as a whole picture.

CHAPTER 11

Coaching & Entrepreneurship– Stepping into Leadership

Georgia was supposed to be a fresh start, and in many ways, it was. After everything, the war, the injuries, the emotional fallout, Georgia gave me space to start healing. Slowly, quietly, One layer at a time.

When I first arrived at Fort Stewart, I couldn't even walk or run properly. The damage to my spine made every movement feel off. Pain followed me everywhere, sharp, nagging, persistent. From the outside, I looked fine. But every step hurt. And no one really knew how much effort it took just to get through the day.

I asked for help, physical therapy, evaluations, but the Army's programs back then were one-size-fits-none. Generic. Incomplete. I'd show up hoping for progress, and leave more frustrated than when I walked in. That's when I met a leader who changed my life. He has asked me to refrain from using his name so out of respect he will remain nameless. He didn't just see rank. He saw *potential*. He saw how I thought, how I worked, how I carried weight that wasn't mine to carry, and he said, "You're going to be a great leader. I'm going to invest in you, not for who you are, but for who you're becoming."

He gave me real responsibility. He elevated me beyond my rank. He assigned me a position that was two levels above my official position and made me his second in command. He trusted my judgement when he couldn't be present. But more than that, he gave me time to heal. Permission to take care of my body and my mind. He allowed me to sign up for civilian education to study physical therapy and from this I ran my own physical therapy program for my own recovery. That belief he had in me was a turning point. And I ran with it.

I started studying physical therapy, not just for a degree, but for a mission. I wanted to turn pain into purpose. I wanted to help others rebuild the way I was trying to. Not just physically, but emotionally, mentally, and spiritually.

I started working with injured soldiers, men who were written off. Who were told they were broken. That they were done. I didn't accept that. I knew what it felt like to be dismissed. I made it my mission to prove the experts wrong who saw no future for these men.

That's where my coaching journey began. At first, it was casual, advice, motivation, basic physical training. But over time, it became something deeper. A system. A philosophy. A calling. Not just about grinding through pain but growing through it. And around that same time, I made a bold move. I bought that first house of mine. Some would consider this to be a common thing that happens in life but not for me. I never owned anything that resembled a home, and I come from a line of poverty. No one in my family owned their own home at the time so it was all a whole new world for me. That house wasn't just shelter for my family and I, it was ownership. A symbol that I could build something beyond the military. Beyond survival. It gave me a new kind of strength, the kind that doesn't come with a uniform. It was my way to say thank you to my wife and kids for sticking around when I was gone and a way to provide a solid life for them.

The man who sold it to me was Eric. He was a real estate agent at the time and a veteran himself. He became a mentor, a friend, and a brother. He helped me long after the sale was over. Treated me with respect I hadn't known how to accept. Over time, our families became close. He opened a brokerage, American Veteran Properties, and my wife became one of his first agents. We helped build the company together, from the ground up. He believed in me. Not because of my past, but because of who I could become. When someone invests in me, I invest back. Always. That's something I hope he learned about me from my actions. Over the years we took turns mentoring each other.

From 2015-2020, I ended up buying 3 different homes. One, when I initially got to Fort Stewart in 2015, another when the Army moved me to Fort Campbell in 2018, and the third, when the Army moved me back to Fort Stewart in 2020. Each time my family and I moved, we rented out our old property and used the VA loan to buy another. This was such an amazing benefit that the military offers, and it really helped me develop a mind for business and ownership. My wife and I used a rental company to manage our very first rental when we moved to Fort Campbell and after a year with the company, we knew we could do better ourselves. They were horrible. They did not manage the property well and never checked on the home to make sure it was being taken care of. After that first lease was up, we knew what we needed to do. We decided that no one would care for our properties the way we did. Even though it was scary, we learned one valuable lesson; Do not be afraid to bet on yourself. The work from managing our properties from there on really showed us that we could do it. We did not always need a company. We could save some money and some headaches if we were willing to do the research and put in the hours. Entrepreneurship became a new extension of the same mission: Building and growing with my family while teaching others to do the same for themselves.

Coaching, real estate, and leadership were all tools to serve the same purpose: healing, helping people stand up again, physically, financially, and emotionally. The deeper I got into leadership, the more I saw how broken some of our people really were. So many started fading. Some fell into addiction. Others took their own lives. And I did everything I could to support them. I picked up the phone. I visited hospitals. I sat with grieving families. I sat in silence with men who had nothing left to say. And it still never felt like enough. Sometimes all you can give is *attention*. Just being present. Sitting in the mud with someone, not to fix them, but to prove they're worth your time. That's why I teach my sons and the kids on the baseball team I recently assistant coached, these three letters: *AEA—Attitude, Effort, Attention.* Because sometimes, showing up *is* the work. And here's the truth: life moves on, whether we're ready or not.

If I wanted to show my sons what true wealth looks like, I knew I had to build it, not with money, but with mindset. With courage. With consistency. I had to show them that the past doesn't get the final word, *you do.* As we continue, examples of how I teach these lessons will bleed through the story by how I show up and stay present for life. We cannot just talk about the actions; we must live them. People will form their own perceptions about us and all we can do is keep living with our values intact. That's the thing about perception: It can shape your world. It's still just an opinion and opinions can change. The way I saw life in my youth... That's not how I see it now. The farther you move from the pain, the more clearly you see the purpose. And I've been building it ever since.

CHAPTER 12

Fatherhood—What You Want to Do Differently

Being a father is the most important job I've ever had. Not the hardest, but the most *important*. Because the stakes are different from war. I'm not just protecting lives anymore; I'm shaping them differently than shaping a young man or young woman. It is just different when you start from birth. Fatherhood changed me. It grounded me. It gave me a purpose that wasn't about surviving the day or completing the mission. It became about legacy. About what I could pass down, not just through money or success, but through mindset, character, and the ability to carry pain without passing it on.

I try to live fatherhood in two lanes: grateful and intentional. I am stern when it comes to teaching my kids what it means to have strong character. I believe in discipline and integrity. In doing the right thing even when it's hard or uncomfortable. I don't let them slack on what matters. I hold the line. Because I know the world isn't soft. And I want them to be ready, not scared, not bitter, but *ready*. For example, I do not allow my kids to answer questions using the word "sure." I want them to be direct when they speak and to hold themselves in a respectable way. I want them to make choices with direct and confident emotion behind them. Saying "sure" to me, just means there is a lack of care on the outcome or the topic of conversation. If I ask my kids if they would like some ice cream, none of them will say "sure." They all would answer with a definite "Yes." I know they are just kids, but the way they communicate will affect the way they move through life, for the rest of their lives. Another way I try to be and show intentional behavior is taking the time to be present and sit with them in an outdoor environment. As people, we have lost the connection we have with the world around us in the belief that the world revolves around us. Occasionally, I take them outside before the sun comes up and we sit in the grass. I teach them to be still and just breathe. To blend in with the environment and see how much they can become a part of the background and not always seek the need to be the center of the world. As the sun rises, they sit and watch. The birds start to take flight, the bugs seem to start moving, the squirrels and grasshoppers start scurrying. It is almost as if the world starts coming to life around you and it was there the whole time. You just have to sit and be still. The wildlife was already there but, in your stillness, you allowed your mind and body to be still long enough to notice. This is teaching them that being still and patient can be intentional, that everything does not require your input and it's necessary to take the moment and see what your surroundings are without letting the "worldly' noise get in the way. One of my favorite things is to watch my children experience this stillness when it lightly rains. I give them a snorkel hooked to a respirator (to allow them to breath underwater). I show them how to sink to the bottom of our pool, and I watch them as they see and hear the raindrops hitting the surface of the water above them. It sounds like nickels tapping a concrete floor. The ripples of the water droplets expand into each other and make the once clear image above a mixed puzzle of what's above. It is such a calming and peaceful moment. I show them these things so they can see the world from different angles and so they can use their mind to see what's around them. When society feels heavy, there is a whole world at their fingertips that will help them make sense of it all. It reminds them that it takes a still mind to think and a steady eye to see. These are great ways to discipline the mind with the body without distraction and even better ways to practice intentional behavior without outside interference from people, social media, opinions, money, success, or reward. Just taking a moment, to be there and be present, intentionally.

I'm also learning to be lighthearted. To laugh with them. Run around the house. Show them that love doesn't have to be loud to be real. That strength and joy doesn't cancel each other out. There is power in grace. Power in forgiveness. As their father I need to show them how to handle failures and successes with grace and how to forgive themselves and others for mistakes. To show them the very human side of being a man while still having healthy boundaries. This only happens through grace and forgiveness. I try to give them both, even when I'm still learning how to give them to myself. I'm not a perfect father. I mess up more than I wish I did. Sometimes I yell. Sometimes I get frustrated. Sometimes I shut down when I should open up. These things just make it hard on everyone involved. The goal is not to be perfect but to show how to grow through the moments we all have that remind us we can do better. I cannot expect grace if I cannot give it and I cannot expect my loved ones to understand if all I do is quietly confuse them and expect more without showing them what more looks like, especially when situations don't make sense to anyone else.

There are times when I step on a toy in the hallway, just a plastic car or a Lego, and my body flashes back to Afghanistan. To the dirt roads. To searching for bombs and landmines with my bare hands. To the hyper-awareness that stepping in the wrong place could cost you everything. When I step on something unexpected, my brain doesn't go to "annoyed dad." It goes to *survival mode.* I try to explain this to my children so they can understand why I might act irrationally but also so they can see me working through something uncomfortable and how I manage through hardship.

My nervous system fires off like there's a bomb underfoot. The immediate response I was trained to have is aggression and violence as I search for what created this threat, but there is nothing to do, nowhere to search, no actions to take. Just nothing. Even still, sometimes I react. I yell. I freeze. I come in sharp when I don't mean to. So, there I am, sweating, fuming, in rage, and I have to swallow it all because why should my family have to deal with my bullshit? They shouldn't. Sometimes it gets the best of me, but I work to explain that everything should have a place to be put away and why it is so important to me that they help me to keep the house in order. The issue is, they're kids. They are young and full of energy and haven't developed mature skills like responsibility. But I explain it. I sit them down and say, "This isn't your fault. You didn't do anything wrong." Because I want them to know that *they're not bad kids just because I have bad moments.*

If I can model that, then maybe they'll learn, struggling doesn't make you broken. That anger doesn't mean you stop loving someone. That healing is real and reachable; it doesn't require perfection. Just honesty. We all do our best, so I hope the times I took to explain my erratic behavior is enough for them to understand that my issues are not theirs, and to give me the chance of not being remembered for my worst days.

I tell them often: "How you do anything is how you'll do everything." The little things matter. How you clean your room. How you speak to your siblings. How you treat people when you're tired or hungry. That's the stuff that builds the foundation for bigger parts of life. I want them to *see* that, day in, and day out. And more than anything, I want them to know: It's OK to not be perfect. But it's *not* OK to stay in pain and call it normal.

If you're not OK, you work on it. You talk about it. You ask for help from trusted sources. You don't let it crush you. You don't ignore it and hope it goes away. We work it out in our house. When we have a bad baseball game filled with errors, or don't do so well on a big academic test. When they seem to be a little more emotional than usual or not enjoying things they usually enjoy, we talk about it. We work through emotions, with a reminder that we are not an accumulation of our mistakes and how we grow to correct our mistakes is building a pattern in life. We go over the fact that emotions are always there and everyone is just one small instant in life from experiencing any emotion. We talk about time and place. If you miss the ball, you finish the play. You gave a commitment to be a member of the team and that commitment does not go away when you get upset or angry. There is a time and place for crying, it is not during the game. I explain to them how to move through emotion so they can come back and experience it later. Emotions are inevitable, but we must hold ourselves accountable for how we react to those emotions. I let them know when they can do better, and they let me know when I can too. And that's not disrespect. That's *trust*. That's what real connection looks like.

I'm not trying to raise perfect kids. I'm trying to raise *resilient, kind, self-aware humans* who know how to get back up when life hits hard. And if they can see me struggle and still *win...* then they'll know it's possible for them too. What more could I ever ask of them?

CHAPTER 13

The Price of Peace— What It Cost to Become the Man You Are

People talk about peace like it's a destination. Like you just arrived there one day, bags packed, demons checked, shortcomings left at the door. But peace isn't a place. It's a process. And it's *expensive*. There's a cost to becoming someone new. There's a cost to breaking generational cycles, to rejecting what you were raised in, to standing in the fire of your past and choosing to walk forward instead of staying where it's familiar. I've paid that price in pieces. In my spine, my nerves, and the pain that never fully goes away. In sleepless nights, flashbacks, overreactions, and moments I can't explain to anyone who wasn't there. In relationships that never seemed to fully recover. In a marriage that became more distant with every deployment, every silent dinner, every word I couldn't find to explain along with the unwillingness to be met halfway. My relationship with my wife, Caitlyn, has been tested over the years. Yes, we made it through the hard times, but it wasn't without sacrifice on both ends.

I've paid it in isolation, because people love the version of you that makes them comfortable, not always the version of  you that's real. I've paid it in growth too. Growth costs you comfort. The willingness to be uncomfortable in life will be the direct reflection of what you get out of life. I found out quickly that I couldn't spend every day with my family if I wanted to lead them through life. I had to be willing to spend time away from them in order to show them a different life. A father and a husband offer much more than financial support to a family. For me, training was necessary. Even though I had to spend time away to get the right training, it does not mean that time was wasted. My family did not like watching me go to training, war, or other countries but they seen me when I was present. They felt my passion and love for them with every moment I got to be with them. My family was beginning to understand when I had to focus on specific moments if I expected to come home in general. We did not enjoy the separation even though we had to respect the time I needed for preparation. I was able to show them how to focus your time and mind on the task and moment that required attention. Even if it was unpleasant.

Peace costs you the freedom to pretend everything is fine when it's not and you confront yourself on your own shit. But I paid it because I *had to*. Because I knew there was more to life than being a disabled veteran in a broken system, trying to act like I was OK while quietly coming apart. My conditions don't define me. My pain doesn't rule me. My pride does not guide me. Peace doesn't just mean no war. Peace means no longer being at war *with yourself.* And for me, it took letting go of who I thought I had to be, so I could become who I was *meant* to be. In prayer. In parenting. In entrepreneurship. In silence. In setbacks. In the gym. Sitting outside in a thunderstorm to watch the trees sway, the rain crash, and the wind rush all to remind yourself how small you can be. In every quiet choice that nobody else sees.

I had to learn that peace doesn't mean perfection. It means alignment. It means walking through your life with integrity, even if you're still limping. It means being able to sit still, look in the mirror, and not feel like a fraud. The truth is, I *am* different now. Because the version of me that went through war... died there. And the version that came back? He had to rebuild himself brick by brick, memory by memory, mistake by mistake. The price of peace is high. It costs time, relationships, old identities, old masks, old habits. It costs comfort but it gives you clarity.

It gives you freedom in the way you see life ahead. It gives you the power to look your kids in the eyes and say: "I fought to be this man. So, you wouldn't have to fight just to survive." Leadership is lonely. Peace isn't soft, It's fierce. It's earned. And if I had to pay the price again?

I would do it; I would do it over and over again. I sat with myself long enough to ask the real questions. Where was I failing? I didn't just sit and feel sorrow and grief. I let go of expectations of others. I let go of the opinions that followed me. I have nothing to prove to anyone, and I don't feel the need to explain that mindset. I just hope everyone reading this can feel if for themselves. I am what I am.

I asked myself if I was being the best father I could be, if I was being the best role model, if I was being the best husband. The answer is no. no to it all. If I sit down and reflect. I can always do more, just a little more effort. It's exactly what I thought when I was facing an impossible task. Because that's what relationships are. They are impossible. Instead of telling others how to act, I am just the man I'm going to be, and I allow them to make the choice to stick around or not. I think about this often when I reflect on the way I talk to the people I care about. Just like everyone else, sometimes I get frustrated in a conversation and I can even hear it in my tone and mannerisms. I immediately know I could have handled the conversation better so instead of focusing on who's wrong or right, I try to change my tone and apologize for getting frustrated. This happens often when I am trying to get the kids ready for baseball practice. They ask their mom where all their cleats, clothes, and equipment is. I immediately get angry and start being aggressive with my tone because the boys know exactly where to put their things and it is their responsibility to keep track of their belongings. Then I take a breath and try to adjust my approach, so I am not scolding them right before they have to go focus on a separate task. This is me working on myself. I believe I have more than I deserve in this life, and I practice gratitude, not anger. I can always do better, and I always will. I won't stop growing and adapting, learning, and loving. Refusing to settle in oneself is the way to live a selfless life. I have weaknesses just like everyone. The difference is that I've made my weaknesses a place of strength and that's the way through.

Who could you be? That is my defining question for the world. Who could you be if you didn't remember yesterday, if you forgot the pain of the past? What could you accomplish if you let go of pointless weight to focus on the way forward? Who would you choose to be if you let yourself just be? Memory is a funny thing.

People often become the people they feared in life. They run away from fear so hard that they stumble into it. We remember to learn, not to dwell. What I have come across in the Military along with life before and after is; most of the people struggling, are the ones that refuse to take accountability for their own actions in life. They refuse to admit anything could have been caused by their choices. They are running from the truth because the truth scares them. It is much easier to blame others and pretend nothing is wrong. I had a close friend a while back that was a good example of this. He would speak on his life as if he had no control of his actions. He would talk about hating his father for being an alcoholic, then continue to drink himself to sleep every night. My friends biggest fear was to be like his father. Instead of choosing to live differently, he saw it as inevitable that he be the man his father was. He would always say "it's in the bloodline" as he joked about his own choice to live at the bottom of a bottle. Instead of making a different choice and trying to avoid the things his father did, my friend leaned into all the things he didn't want to be. Trying to protect himself from the thought that it was still his choice to continue to drink. His biggest fear was being like his father and yet, he did the same things his father did, but that fear is not reality. Confronting himself would be difficult but it could be the start of his own peace.

Peace costs more than you can afford and yet, you must

find a way to pay the price so you can grow the world around yourself. My legacy will never be my name on billboards or in movies. It will never be a worldwide name to remember, and I relish the idea that I will be gone but the lessons I have given to those around me will live for generations to come. My name does not need to be remembered for my ripple in this world to build momentum.

CHAPTER 14

Bodybuilding

In 2022, I was placed in a different role, overseeing the Career Counseling and retention program at Hunter Army Airfield in Savannah, Georgia. Earlier, in 2016, I had spent some time at Hunter teaching breaching techniques to some of the breachers of Seventy-fifth Ranger Regiment. It was a great time of my career, and I missed that, so I was excited to go back to a familiar area.

Every year, the Headquarters of the Army hands out certain numbers of personnel we need to sustain in order to have a fully staffed military. In September 2022, my team in Hunter completed the yearly mission that the Army set for us. When I took over, we were one of the worst units on paper but Vassey and I, along with a new counselor, SSG Browning, ended up helping that number that was given to the Unit at the beginning of the year. This means we helped these soldiers ask and receive what they wanted from the Army. Some wanted to change jobs, some wanted to go to a new duty station, and others just wanted to stay where they were because they liked the school district for their kids.

Even though there is a goal number, these are people with real life needs and aspirations and my team understood that.

We did not sacrifice our care and efforts to maintain good numbers. In fact, that is the only reason we succeeded. The soldiers knew they could trust us, and we would help them even if it didn't help us. Once we accomplished what some would call an impossible mission, I got busy training for bodybuilding competition that October.

A friend of mine had had surgery and wanted to compete as a bodybuilder after his recovery. I had never been a huge fan of the sport, and I knew so little about it. The more I learned, the more I respected the athletes were disciplined and worked hard to develop their bodies in these certain ways. I told my friend I would train and compete with him, and we could hold each other accountable. It was my way to show him support in a hard time. If not for him, I never would have done it.

When I got serious with my prep in July, I had a solid foundation for bodybuilding just because I had always worked so hard to maintain my physical fitness. But everything seemed to come at once. During this time, my friend moved so I was on my own in training, and it was also another busy time with work at Hunter. My family was moving into a bigger home, so that we could turn our previous home into a rental. I ended up becoming extremely sick, with a fever of 104 degrees. But with tenants waiting to move in, we still had to get all our stuff out of the old house and into the new one. As if my plate wasn't full enough, a broken waterline flooded the new house. We had to replace all the flooring downstairs and order all new appliances, which took forever to arrive because of lingering supply-chain problems leftover from COVID. For six weeks, my family had no kitchen appliances to cook on or store food. But nothing good comes easily.

Once everything was taken care of at home and at work, I took some much-needed leave and focused solely on the bodybuilding competition. It turned out I had enough time and drive to make it work! I took first place in my first and second categories. Without my coach Evan, this never would have been possible. My children were watching me on Vassey's TV while I received my two first-place medals. It was the first time I felt that I got to share a victory with my children. Explaining success in the Army or related topics can be hard to communicate to a young child, but this was something they could easily understand, especially when they got to see the award ceremony.

My wife, Caitlyn, was also a huge reason I won. For starters, she did all the paperwork to sign me up. As she was crushing her own career in real estate, she was also prepping and cooking all my meals. The competition was a four-hour drive from our home; she handled all the logistics and drove so I could keep my cortisol levels low and focus on the competition. I never would have done as well as I did without her help. It was another win for my children to see us work together as a team to accomplish a goal.

CHAPTER 15

Trading Comfort for Calling—The Southern Gym Supply Gamble

Money never motivated me the way it seems to motivate most people. Don't get me wrong. I like security. I like being able to provide. I like not flinching when bills show up or when my kids need something. But the goal was never wealth. It was *freedom*. And more than that—it was *impact*. For a long time, I thought I could find that through real estate. I built up a small portfolio of rental properties, nothing massive, but solid. Safe. The kind of thing people nod approvingly at when you say it out loud. It looked good on paper. It made sense. And for a while, it worked. It gave me a steady income and some breathing room. A kind of controlled success. But deep down, I felt it. That quiet itch. That *"this isn't it"* feeling. I was surviving, not building. I was maintaining, but I wasn't chasing anything anymore. And if you've read this far, you already know, I'm not built to sit still. So, when the opportunity came to step into something bigger, something aligned with who I've become and what I believe in, I didn't hesitate long. I sold off some of those properties. Equity for purpose. Comfort for risk.

That's when I met Leon Anderson. Leon wasn't just some fitness guy with a logo and a dream. He was a builder. A man who knew the value of tools, grit, and relationships. A quiet leader. Someone who had carved out a niche in the shadows of the franchise giants. He believed in giving people the blueprint to build their own gyms, on their own terms. No contracts. No corporate red tape. Just real equipment, real strategy, and real freedom. My original plan was to build a gym of my own, but after some thought and planning I decided to become a partner in Southern Gym Supply. That partnership didn't just change my business life. It changed the way I seen building and manufacturing. The idea of helping someone bring their dream to life was something I wanted to experience for myself. That was a powerful thought for me and there's something sacred about that. Especially for people like me who came from little. Who had to fight for every inch. Who weren't handed blueprints, but figured it out through fire, grit and failure. I am not just talking about finances. I am referring to mentality.

Being able to build something to run and operate the way you see fit. Caring about the people and community that are using the equipment you've spent time and energy constructing. Finally, taking that step to create something privately owned instead of a corporate approach. We are not here to get as much money as possible regardless of actual effect. We were there to build a new type of freedom. I believed in my message so much, I chose to show people what it could be, all they had to do was take a chance on it. Bet on themselves and live full. Funny thing is, I was never chasing success or money. I was chasing freedom, and that's the life I am able to live now. I am just trying to help others reach the same gift. I recently helped a good friend of mine build a 10,000 square foot gym and it's going great. He is learning what true freedom feels like. Growing his own legacy, his children can grow from. Southern Gym Supply brought me more business and like-minded individuals. Well, close to like-minded as they can be. I can say without a doubt, I've never met another man like me. Not good or bad, just the truth.

CHAPTER 16

Iron House Athletic Club

After the Bodybuilding Competition, several friends and strangers reached out to ask for my guidance and mentorship in building strength and physique. I had spent over a decade training and conditioning soldiers, so this was right up my alley. Different, but definitely in my wheelhouse. I started coaching and my clients' results spoke for themselves. From there I decided to find my own team and open my own gym. I already owned half of a company that built commercial gyms for private clients, so I decided I would build my own with the same process I built for others.

I used Southern Gym Supply to build my gym just to show people that it was worth it. At the time, I built Iron House Athletic Club to be functional but to also be a showroom for Southern Gym Supply, to show people my custom equipment and be able to demonstrate why that equipment was the best around. I didn't even care if the gym made a profit. I just wanted it to come out even and I would have been happy.

Little did I know, the community felt that message in their bones. People came from other towns just to see the gym. We really built something amazing, and it was just a 3,200 square foot gym, quite small for a commercial gym. Everything there was structured to save space, and the layout made it seem that there was more space than there really was.

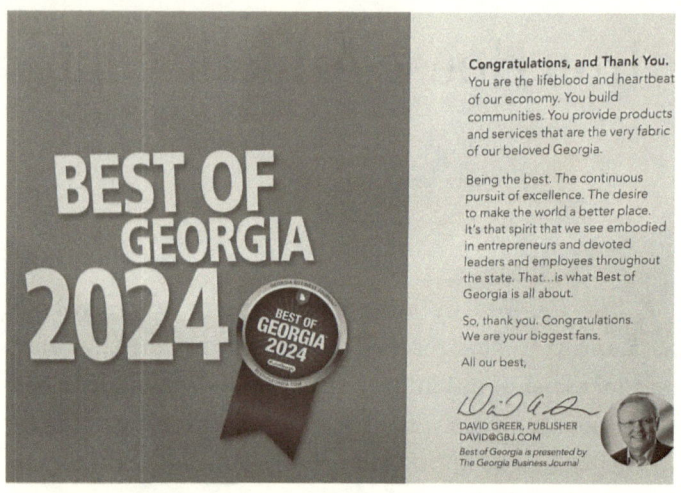

Congratulations, and Thank You.
You are the lifeblood and heartbeat of our economy. You build communities. You provide products and services that are the very fabric of our beloved Georgia.

Being the best. The continuous pursuit of excellence. The desire to make the world a better place. It's that spirit that we see embodied in entrepreneurs and devoted leaders and employees throughout the state. That...is what Best of Georgia is all about.

So, thank you. Congratulations. We are your biggest fans.

All our best,

DAVID GREER, PUBLISHER
DAVID@GBJ.COM
Best of Georgia is presented by The Georgia Business Journal

There's something sacred about building something with your bare hands. From building walls and laying flooring to running wiring for doors to close just right and painting. Not just the work or sweat, but the soul you pour into it. For years, I built gyms for other people, private owners chasing their dreams, corporate investors opening facilities, trainers looking to level up. I delivered quality, I delivered results. But deep down, I knew: One day, I'd build one for me.

That dream became real in Hinesville, Georgia. Iron House Athletic Club wasn't just a business; it was the next mission. I was transitioning out of the military. The uniform was coming off, and for the first time in decades, I had to ask myself: What now? I knew the answer had to include purpose. Growth. Structure. Leadership. I couldn't sit still. I couldn't fade out. So, I leaned on what I knew, discipline, training, and a vision that had been simmering in my gut for years. I gave myself ninety days. No investors. No big corporate backers. Just me, my company, Southern Gym Supply, and a mountain of work.

Caitlyn, my children, and I handled everything in-house. The equipment, the flooring, the layout, the software systems. I designed it to be 24/7, raw and real, a place where bodybuilders and powerlifters could train without frills, without distractions, and without compromise. Iron House wasn't about polished chrome or treadmills with TVs. It was about iron, grit, and growth. My children helped me lay the rubber flooring; they watched me build the walls and doors. They got to see me building and watch their mom, my wife, help. It was an amazing family growing activity all around. It was a vision coming to life in front of my family and unlike war, I could include them. They got to see the work and be included.

There were no employees. Just two independent contractors: and a sports massage therapist. Everyone had their lane, and they handled it like pros. I didn't need a huge staff—I needed a strong system. One built with the same philosophy I gave to every client I ever served: build a foundation so solid it doesn't need constant supervision.

The gym paid for itself faster than I imagined. Within the first month, it was turning a profit, not from flashy ads, but from community trust. I coached and mentored athletes on my own time, and that revenue helped fund the supplies and final equipment installs. I reinvested every dollar with intent.

Here's the wild part: within ninety days of opening, I wasn't tied to the front desk. I wasn't sweeping floors at 2 a.m. I had already built it to run itself. I was taking trips, visiting family, and spending time with my kids, not because I left the mission behind, but because I had executed it. The sign-up system could be scanned from the outside and the lock on the front door opens via Bluetooth on a person's phone once they signed up. That kind of freedom doesn't come easy in business. Most entrepreneurs burn themselves out trying to be everything for everyone. But I focused on building the right systems, the right team, and the right energy. And the community noticed.

In the same year we opened, Iron House Athletic Club was voted "Best in the Region" by the *Georgia Business Journal*. Not because we paid for exposure, but because people believed in what we were doing. They felt the energy, the authenticity. They walked through the doors and knew they weren't just in a gym; they were in a place built with intention and love.

My kids were able to watch the entire journey unfold. They saw the 6 a.m. truck deliveries, the 10 p.m. installs, the tired eyes, and the full heart. They saw that success doesn't just show up. It's earned, rep by rep, hour by hour, decision by decision. They also learned that success comes in many forms—not just in medals or bank accounts, but in impact, independence, and legacy. Like I said: Iron House wasn't just a gym. It was the first step into life after the military. It was proof that I could adapt. That I could lead without a rank on my chest. That I could build something bigger than me and trust it to grow. And the real lesson? It all came back to the same thing, build the team, build the culture, lead with character, and success will follow. I built Iron House. But in the process, Iron House built me too.

Now, I am currently in a position to where I will never need a "job" for the rest of my life and it all comes full circle. When I was young, the gym gave me a home. I shared that with others, and it brought me to a place where I will no longer have to do anything someone else tells me to do. I can live a life growing and molding my kids.

I am thirty-four years old, and I've been retired for over a year now. Funny thing is, I was never chasing success or money. I was chasing freedom, and that's the life I am able to live now. I am just trying to help others reach the same gift.

CHAPTER 17

Retirement and What's to Follow

My retirement was a little unconventional for the Armed Forces. I was given an early retirement, so I did not serve the whole twenty years. I was moving the whole time, though. It seemed like there were no breaks and I was just trucking through. I was a master sergeant in twelve years. The military is not a race for rank, but it was nice that the Army saw value in the efforts I gave and kept rewarding them. I often sat in positions well above my pay grade even though I wasn't getting paid to do so. Most military personnel know what that's like: More responsibility and more work with no promotion and no pay raise.

It's not always a great feeling but to lead is a gift and if you don't see it that way, you should rethink the way you are moving forward in life. For people to give you their lives and trust you to make the best choices for them and those around, that's a massive weight that should be fostered and respected from all angles.

My promotion to master sergeant was my last promotion in the Army. I didn't invite anyone; I usually move in silence. But my Command and the Division Career Counselor team

sent out invitations, asking if anyone wanted to show up. Many more people than I expected were there. It was the first promotion my family attended, my children, my wife, and my stepfather Troy. It was a great honor to me that they took the time to be there with me. I was happy that my family got to see that moment; I was happy to be part of an example of what it looks like when people show up for you out of respect and not obligation.

I will now share the speeches of my two guest speakers as well as my speech at this ceremony. First up we have Darrell Vassey's speech.

Good morning, everyone: family, friends, distinguished guests. I'm Sergeant First Class Vassey. I'm a good friend of Master Sergeant Rowe. It's always a big honor to speak at his promotion on his behalf, so it's always an honor when a friend, colleague, and mentor asks you to speak on their behalf of their promotion ceremony. This is a big day for him.

I'm trying not to let him down, so I wrote a bunch of stuff down. I've known Master Sergeant Rowe for a long time. We met in 2016 at Combat Engineer Advanced Leaders Course at Fort Leonard Wood. We got to know each other, we became good friends, and then we went our respective ways after ALC.

Fast-forward three years. I'm stationed at Fort Campbell, and I get a call from somebody, and he says, "Hey brother I'm coming to Campbell." We spent the better part of a few years at Fort Campbell together and then I came here to Fort Stewart and guess who comes right to rock of the Marne after me? Master Sergeant Rowe. So, we've been spending the bulk of our career together, friends, colleagues, mentors, working together side-by-side. I couldn't be happier to get to work with my friend every day.

When I think about Rowe as a soldier, a husband, father and a friend, a few words come to mind. The first word is loyalty. Jacob is fiercely loyal to those he cares about. He'll give you every bit of his time and energy even while he's pursuing his own goals. Next thing is tenacity. Jacob is always leading by example. He's always there to share a life lesson with you or somewhere in the weeds, taking care of you as his soldier or his command team and bettering the organization, himself, and those around him. The last word that I can think of for Jacob is character. Jacob has never been an individual that I had to question his character or his moral compass. They do not come more straight up than this man. He is direct and truly compassionate, and everything he does is the better himself, his family and those around him, with that I will turn it over to Master Sergeant Jacob Rowe. Congratulations on your accomplishments and everything that you have done; you have earned it.

Next, we have Master Sergeant Bernard.

Hi! So, you all don't know me, but I am Master Sergeant Bernard, retired. As you can see if you could not tell the full-grown beard. I came here in support of Jacob because he became a friend, but I met him when I first got here 2018 from the schoolhouse. I was an instructor for Senior Leader Course for about three years and came to Fort Stewart to become a senior for a brigade. I did not know how I was going to look like in real life because in school you teach and then you've got to come to that real world. I did not know what to expect so I was kind of scared going into becoming a brigade senior.

Jacob was the Division Operations NCO here. That is when I first met him. I will tell you, when you are looking for a leader, they do not have to be above as a superior. There are subordinate leaders out there who will shine and perform when it is needed. Jacob was one of those leaders for me coming in as a senior. I did not know what was going on, so I went to him, and we had a lot of long talks about the things that I needed to learn and catch up on. He learned the seniors and the office and we kind of just bounced off information from each other. So, I learned a lot from Jacob, which helped me become a better senior. Also, the team that I had was an amazing team, but just to have him at the division operations level, it helped me become better at what I needed to do.

There are a lot of conversations that Jacob and I had that made me grow as a senior and as a person because I got to know the personal side of Jacob and who he was as a person, not just a soldier, but as a leader. So, I will tell you if you're looking for a mentor, this is the guy that you need to find. Many times, we sat down and went through choices and decisions to challenge each other to grow into our leadership roles. He is a good friend. He is a good leader in what he's doing for the community.

If anybody knows, he is opening a gym in Hinesville. It's amazing what he's doing being able to help everyone now. I'm not disciplined enough because I love food and so I tried it on and it just didn't work for me, because you know food be calling me but if you need anything from him as far as leadership advice, financial advice, fitness advice, or personal advice, he can give it. His wife is in real estate, and I work with her as well.

He's just an amazing person, so I can't say anything but good things about Jacob and this is a well-deserved promotion. He's already been filling in in the position of a senior because he has already been senior in the past and so this is just long overdue. Now they put the rank on there, and he can finally get the money out of it too. Congratulations well done."

Last, my speech.

"There are just a few a few things I want to hit really quick. If you know me or you've met me, you know that if you give me a floor, I'm going to speak. So, in any kind of promotion and how I see the Army as a whole, anyone who gets promoted, the promotion isn't for you. The promotion is for the effect you have on the people around you. For what you can bring to the organization by having a bigger role. Any promotion for you means more responsibility, which, if you understand responsibility, it's a curse.

If you do something great, it's because of your subordinates' efforts that made it right. If I do something good, it's because of people around me. If something fails, it's on me because I'm the leader in the scenario. My responsibility. There's a difference between two things and I talk to people often. I talk to hundreds of people a day, and one question I always ask is "What do you want?" I think as any individual or any organization is trying to grow, you need to know what the people around you care about and where they want to go in life. It's less about what you want and more about what your impact is. When you're talking about somebody's life, what they want out of life hits different from when you look at their impact on the world around them.

The needs and wants of an individual change vastly based on the impact that is made while serving in a certain position. So, ranks and structure really don't mean anything. It's the respect of the position that you hold. I deserve nothing from everyone here, but I'm really grateful and I appreciate everyone being here. I don't deserve it. I may have earned a position in the Army, but the respect comes with the position and what you're expected to hold while you're in it.

Second, I want to speak on problems with no solutions. If you can bring a lot of problems to the table in any organization, you're at, without having solutions, You're the problem. If you have never thought about the solution to these problems, all you're doing is going around and popping bubbles and or pointing out holes when you should be brainstorming ways to mend the issues. Any problem is an opportunity to find an answer. If you don't have an answer for the problem, your mind is going to go toward what you're focused on instead of what you should be looking for. If I walk around looking for problems, I'm going to find problems. If I walk around looking for character and good people, that is exactly what I will find. If I walk around looking for horror and tragedies, that's all I'm going to see. So really, it's a perspective. How you live your life in perspective is going to dictate who you are as an individual and what you find yourself invested in.

Lastly, we have, motivation. I don't care about it. It's useless. It lies to you. It doesn't mean anything. I had no motivation to get promoted; it happened because I was disciplined in my career and in my circles; I was disciplined in the things that I needed to do. I don't wake up every day and feel motivated. Feeling motivated is having a feeling to want to do something. I don't always want to go work out. But I do that shit anyway because I know it needs to be done. I don't always want to eat healthy food. But I know I like the way my body and mind operate when I do. So, I shut my mouth and eat that shit anyway. If you got everything you wanted, you'd be in a personal prison that you've built yourself. It's discipline over motivation. Motivation is fake and discipline is real. Just understand this, in leading, motivating those you lead is extremely important. Finding what motivates your people is a great way to show that you care for them. Being a source of motivation for others is an amazing leadership quality. I am simply saying, discipline is necessary regardless of how we feel as leaders. We are here to motivate, not to rely on that feeling to push us through the impossible tasks of leadership.

I just wanted to say those things because I think it's important to hear every once in a while. I would like to say thank you for everyone coming, specially to my family. This is the first time my family got to join me in something like this. I appreciate you all. I apologize for not sending out personal invites, but it just means that much more that everyone's here so thank you so much for coming."

Retirement for me, made sense. I should be tired. After everything, after war, surgeries, broken bones, sleepless nights, rebuilding a family, launching businesses, losing friends, letting go of the past, and dragging myself through grief, you'd think I'd want to slow down. And yeah, there were days I told myself I earned that. *Take it easy, you've survived enough, you've done your part,* but here's the truth:

I'm not done.

As long as God gives me breath, I must continue to honor and cherish the life I am given. There's a different kind of fire that shows up after you've lived through the worst. It's not about proving anything to anyone. It's not about attention. It's not about chasing some title or number or accolade. It's about refusing to die with work left in you. That's what drives me now. Not my past. not rage, not fear, *Purpose.*

I wake up every day with this quiet urgency in my gut—not panic, but clarity. I know how fast it can all be taken. I know how short life can be and I know what it feels like to waste time. Now I'm building, I'm serving my family, I'm *working* on myself, on my mission, on the man I still want to become. The finish line isn't survival. The finish line is *impact.* There's a misconception people have. They think once you "heal" or "find peace," you stop pushing. You stop chasing something bigger. You sit on the porch and exhale. I don't want peace that looks like passivity. I want peace that fuels movement. I don't want calm that makes me soft. I want calm that makes me *focused.* The only difference is that I am not reckless with my time and intent anymore. My focus is not wild and untamed like it was when I was younger. It's controlled now. It's deliberate. It's aimed and it's relentless.

I look at my sons and I think: *They're watching,* they're watching how I carry pain. How I talk to myself when no one's listening. How I react when things fall apart. They're watching how I get back up. What I chase. Who I serve. What I protect. What I stand for. I can't ask them to live with courage if I live in comfort. I can't ask them to chase their potential if I settle for mine. I can't ask them to be disciplined if all I do is talk and never *show.* So, I lead from the front. Even when I'm tired, especially when I'm tired.

Because love isn't soft. Sometimes love looks like sweat. Sometimes love sounds like doing it again but better.

People ask me sometimes, "What keeps you going?"

It's not one thing. It's all of it. It's the kid I used to be, the one who thought he'd never make it out. It's the soldiers I buried, who never got the chance to build what I'm building now. It's the broken version of me in that hospital bed, wondering if I'd ever be strong again. It's my wife, my sons, my family, and every client who looks to me for direction and expects me to *show up*.

There's a version of me I haven't met yet. A stronger man. A wiser one. More grounded. More generous. More patient. More present. A man who's not led by wounds, but by wisdom. A man who still walks with fire but no longer uses it to burn bridges.

That's who I'm chasing. Not to impress anyone. I just don't want to become a man who lets time pass without seeing it as the opportunity to have new life every single day. I'm not looking for breaks. I'm looking for *breakthroughs*. Let this chapter be a warning, a promise: *as long as I have life, I will do my best to honor it.* Because one day, will be my last. One day, time will call my name for the last time. And when that day comes, I want to be able to say: "I used every bit of fuel I was given. I didn't waste the pain. I didn't hide from the pressure. I didn't coast on what I did ten years ago. I gave the next generation a blueprint, not just words."

That's what the time I have left is for. Not ego. *Legacy*. And as long as I'm breathing, I'll keep building it. One day, one rep, one choice at a time. Because if I draw another breath, I still have work to do. Legacy isn't something you declare. It's not a brand. It's not a title. It's not a plaque on the wall or a bio someone reads at your funeral.

Legacy is built day by day, action by action, choice by choice. And the truth is, you don't get to claim it. The people you lead decide if you lived it. We live in a world obsessed with noise. People shouting their wins from the rooftops. Every lift. Every deal. Every vacation. Every "look at me" moment.

But the men I respect? They move in silence. We do the work when no one's clapping. We carry the weight no one else can see. We win in private, and we celebrate those wins with the people who *earned* them with us. Legacy isn't built in the spotlight. It's built in the dark. In 3 a.m. wakeups. In late-night phone calls. In the moments you hold your tongue when you want to explode. In teaching your kids the same values you're still mastering, and it's not for you.

Let me say that again: *your Legacy isn't for you.* It's not about being remembered. It's about being *followed.*

Legacy is what happens when your kids hear your voice in their head telling them to get up, to finish, to lead with character, even when you're not there. It's the way your team keeps the standard high when you're not in the room. It's the way a client teaches their son how to lift because of the discipline you instilled in *them.* It's not a statue. It's a ripple. Quiet but deep.

Some people think legacy is about status. A name in lights. A "look what I built" kind of life. But I've seen what real legacy looks like. It looks like my son opening the gym door for a stranger, because *he watches how I treat people.* It looks like a friend repeating something I said three years ago because it stuck when I didn't even know they were listening. It looks like a man showing up to a funeral to say nothing other than to show presence. That's the stuff that lasts. That's the stuff that echoes. You don't get to schedule it or measure it. You just have to live in a way that makes it inevitable.

When I think about legacy now, I don't think about money. I think about moments. The conversations that shape someone's direction. The choices my sons will make because of the way I handled adversity. The respect my wife gives me not because I demanded it, but because I earned it. The strangers I'll never meet whose lives are better because someone I led... led *them*.

Legacy is leadership that outlives you. And the only way to build it? Is to lead now. Not when it's convenient. Not when you're famous. Not when you "feel like it." Now. Every single day. Because the truth is, someone's watching. Someone's learning. Someone's deciding who they're going to become based on how you move through this world.

That's why I don't chase validation. That's why I don't need everyone to "get it." That's why I don't argue with small minds or beg for support. I know what I'm building.

And I know who I'm building it for. My legacy isn't a monument. It's a mirror. And I want the people who follow me to look into it and see their *own greatness*, not mine. Because that's what real leaders do. They build others. They multiply strength. They make their ceiling someone else's starting line.

So, I'll keep moving in silence. I'll keep building when no one's watching. I'll keep showing up, even when it's inconvenient, even when I'm misunderstood. Because legacy isn't claimed. It's not something you demand.

CHAPTER 18

Show Up in Life

Showing up in life is something that life constantly demands of us. Most of the time, people proclaim their good intentions or the fact that they tried or wanted to be there for the people they care about. The truth is you make time for the people you truly care about. When my mom graduated from school with her master's degree, I made time to travel from Tennessee with my family to attend her graduation in Ohio. I've spent thousands of dollars to make sure my mother had food and shelter and to support my mom by showing her that the people around her cared about what she had to go through to accomplish the things she has achieved. In her own story, she should've been a statistic, but she wasn't. In my eyes, that made her graduation something important to show up for.

Same thing for my little sisters. When my sister Hannah graduated with her degree, I showed up for that as well. But "showing up" isn't just about being there for the good times and the celebrations. It's also about showing up for the hard times and the bad times. When my sister moved to her apartment from my mom's house, I traveled from Georgia to Ohio to help.

In 2021, my grandmother and aunt were in a bad car accident right after they had to move out of their house. They both sustained major injuries. Again, I left Georgia to show up. I went to the hospital to take my grandma home. I walked my aunt and my grandma into their new home while they were still only half moved. Some of my family lived in the same town as them, while others didn't. I was pretty upset with the fact that I was the only one there. There were different times where other people showed up, but when it was time to get to work, help them move their house and make sure they were OK, I felt alone. I was the one in the middle of the night, moving their stuff from one house to the other. I was the one driving them around and ensuring they'd be OK.

I love my family, but it's also important to open your eyes and realize that when you are the only one showing up, you need to start setting boundaries. I wasn't going to not help my eighty-year-old grandma. But I was definitely going to tell the people who should've been there helping that they needed to show the fuck up and I couldn't keep being the only one.

Right after this, I drove four hours to my mom's house to help her and her husband move from Ohio to Atlanta, Georgia. I helped them pack up everything in their third-floor apartment, carry it downstairs, get it safely in the moving van, and then drive to Atlanta. I did this with no sleep and no rest and so much to do. After that, I drove back to my house in Georgia, but at least that was only four hours away.

Recently, my mom moved from Atlanta to Cypress, Texas, so my mother could start her job at MD Anderson Cancer Center. While most of the heavy lifting was done by her husband and hired movers, I still showed up to drive the truck.

My early years in Georgia were pretty hectic. There was a lot going on and I had just got home from a thirty-day training rotation in Fort Polk, Louisiana. These training simulations are to practice logistics of counter, moving parts as well as combat operations and to test the maneuverability of units. I wasn't being tested, I was helping run live fire exercises. So around sixty to seventy live fire missions. That means each mission was three rotations: one dry (walkthrough), one with blanks (practice for live ammunition), and one with live ammunitions.

I was teaching briefing techniques to the new engineers and showing them how to use their demolitions. Every time I had a team go, I went with them. We split into two teams so half of our group would get rest, and the other half would go to work. I went on every single one because I was responsible for the accountability and special equipment of my team. I was responsible for my people and the demolition. I facilitated my team leaders by showing up for them and showing them support but giving them full control of their team because I trusted them. I just waited in the back and helped them make decisions if they had questions. I empowered them to be leaders. I showed them that I would still walk the ten miles next to them just so they felt like they had someone to turn to if they needed it. I've done a bunch of Joint Readiness Training Center (JRTC) and National Training Center (NTC) rotations. These rotations are usually four to six weeks long. In California for NTC and Louisiana for JRTC. It's hectic and throws wrench in any plans you have for you and your family, but it needs to be done. After one of these rotations, as soon as I got home, I learned that one of my brothers had shot another brother in the chest with a pistol. I didn't have time to spend with my kids. They were all right and taken care of with their mother at home; they didn't need my immediate assistance. So, I showed up for my mom and my siblings. I drove from Georgia to Ohio to ensure that my brother was going to be OK and that he knew he had support.

Of course, I realize that people can't just stop their lives and show up for you simply because you have some big thing going on. There's a level of understanding we all need to have with our loved ones. But in my life, I allowed people to not show up for me by saying nothing about it when it happened. I did not want to burden my loved ones with my life. That is just how I felt about it. More like a burden because I taught them how to treat me by not expecting anything from anyone. So, in some sort of inadvertent way, I was responsible for my situation. Because I never said anything and because I didn't communicate properly, I gave them permission to not show up, but because I tried to show up, they learned to expect more from me than they were willing to give.

My mom and Troy have shown up far more than anyone else. Even Bob showed up to help me move to Georgia the first time. But when we go through it all, countless graduations with no one there, business adventures, and grand openings, my wedding and much more. I'm sure there's a good reason. Now looking back, it's just awakening. But at every point, you are showing up for yourself too. You are showing up in the world to be present in that specific moment. You are showing up so you can be the man you say you are. You show up because it matters. Because at the end of the day, if you lived your whole life just for you, you've lived a very empty and lonely life. Some of my fondest memories are some of the hardest relationships and hardest memories

I've ever had. I show myself who I am every single day with every single choice.

CHAPTER 19

The People You Fight For

Relationships are strange. You'd think the people who've known you the longest would be the ones who show up when it counts. That loyalty would be automatic. That history would mean something. But I've learned the hard way: It doesn't always work that way. Some of the people I'd take a bullet for barely check in. And some of the people I've just met? They show up like family. That used to bother me. Now I take it as truth. Because life's not about who *knows* you, it's about who *values* you.

In the military, relationships are built fast and deep. It's not surface-level. You're depending on the guy next to you to keep you alive. You sleep near each other, eat together, bleed together. There's no room for masks or small talk when death is around the corner. You learn who people are by how they act under pressure. And that kind of brotherhood is real—but it's also *contextual*. Because when the war ends, and the uniform comes off, and there's no mission tying you together anymore, something changes. Sometimes, you drift. Sometimes, it's silent. No falling out. Just a slow fade.

No one wants to admit it, but proximity built that bond more than compatibility. Some of those guys are still like blood to me. Others? We haven't spoken in years. And not out of hate, just... distance. Life. On the other hand, business introduced me to a whole different kind of relationship. Some of my clients have become better friends than people

I've known for decades. Why? Because they *see* me. Not just my resume. Not just the past. But the effort. The intention. The way I show up. The way I help without asking for recognition. And when you help someone change their life, build a gym, start a brand, transform their body or mindset, that gratitude turns into real connection.

Some of these folks I met on a sales call. Now I know their families. Their birthdays. I've stood in their gyms and watched their vision become real. It's wild. The people who paid me ended up treating me better than some who grew up with me. Here's the hardest part to admit: Some of the people closest to me, family, childhood friends, people I would've given the shirt off my back to, they've taken me for granted. They don't see the man I became. They see the kid I used to be. Or worse, they see a version of me that fits *their* comfort. I'll be honest, there were moments that crushed me. I'd look around at a room full of people I've supported through hell, and not one of them clapped when it was my turn. Not one phone call when I launched a new business. Not one text when they knew I was hurting. But I've learned this: *Don't expect people to celebrate a version of you they never had the courage to become.*

Some people stay small. Some people only love you when you're struggling because your growth threatens their stagnation. It's not hate, it's fear. But it's not your job to shrink to make others comfortable. In business, loyalty is a gamble too. I've been burned. People I mentored have stabbed me in the back when it suited them. People I vouched for disappeared when it got hard. Clients I bent over backward for ghosted me the second someone offered them $20 off. But I've also had people ride for me in ways I didn't expect. Clients-turned-friends who've called just to check in. Business partners who've prayed over me. Strangers who became brothers because they believed in the mission before it even had traction. So, I've stopped assuming loyalty comes from time.

Now I look for *alignment:* Do our values match? Do you move with integrity when no one's watching? Do you reciprocate or just take? Relationships will test you, especially when you grow. Some people don't know how to love you once you heal. Once you stop being the fixer. Once you put up boundaries. Once you become the version of yourself, they didn't expect, or can't control. That's when the real ones show themselves. And I'll say this: I'd rather have three people I trust in a fire than a hundred people who liked my posts and forgot my name. So, if you're building something, business, body, brand, life, look around. Who claps when you win? Who leans in when you're low? Who calls, not to ask for something, but to *offer* something? That's your tribe. Not the longest friendships. Not the loudest ones. The *true* ones. And when you find them? Pour into them. Protect them. Partner with them. Pray for them. Because the people you surround yourself with… They shape the kind of man you're allowed to be. And the older I get, the more I realize, I don't need more followers. I need *fewer, better* friends.

CHAPTER 20

A Message to My Kids and to You

To anyone reading this who feels like they've lived a life too heavy to carry,

I wrote this book, so you'd know the *truth*. Not the polished version. Not the highlight reel. The real one. The one where things hurt, but healing still happens. The one where scars don't disqualify you, they *qualify* you to lead, to teach, to become something greater.

If you take nothing else from my story, take this:

You are not stuck.

You are not broken.

And you are never too far gone.

Life will come at you hard. It will test you, stretch you, take things from you that you thought you couldn't live without. But if you keep moving, if you keep choosing growth over comfort, purpose over pity, and faith over fear, you *will* build something meaningful.

To my sons:

You were my reason when I had none. You made me show up when I wanted to shut down. I wasn't always gentle. I wasn't always right. But I was *always trying*. And I promise you, I'll keep trying. Every day.

If you ever wonder why I'm intense about certain things, why I care so much about integrity, or why I raise my voice when the floor is a mess, it's not because you failed. It's because I've lived a life where a single misstep could have ended everything. And that stays with you. It wires your instincts. But I'll always explain it. Not because I owe you perfection, but because I owe you *honesty*. You are not responsible for my past. But you *are* part of my purpose. And

I want you to know that even if I fall short sometimes, my love for you never will. I hope you learn that strength isn't how hard you hit, it's how well you carry what hurts. That forgiveness is a weapon, not a weakness. That it's OK to feel, just don't let your feelings run the whole show. And above all, I hope you see me struggling and still *winning*. Not to impress you. But to remind you: if I can, *you can*. To anyone reading this, maybe you didn't grow up like me.

Maybe you did. But if you've ever felt invisible, underestimated, or undone by life, I want you to know this:

You can become more. More than your hardships, more than your upbringing, more than what people think of you, more than who you were yesterday. Remember, it will cost you. You'll lose people. You'll lose comfort. But what you'll *gain is* peace, clarity, self-respect, and that's *everything*.

We live in a world obsessed with shortcuts. With likes, with followers, but none of that means anything if you can't look yourself in the mirror and respect the person staring back. This book wasn't written to make me a hero. It was written to remind you that *ordinary people survive extraordinary things* every single day. And sometimes, surviving is only the beginning.

The rest of the story, the one that really matters, is how you rise. Even if it's slow. Even if it's ugly. Even if no one's clapping for you yet. Keep going. Because how you do anything... is how you'll do everything. This world needs what only *you* can bring.

CHAPTER 21

Now—Building Peace, Living Purpose

Life looks different now. Not just for me but for my whole family. My brother Zach has been sober and living well for over a year now, spending his time with his kids and forging a life for them. Zachary healed after being thrown out of a window, but there was still some permanent damage. His voice will never go back to the way it was before the accident. His brain operates a little different since then. The scars on his head will always be there. Even through all of this, it didn't mean he couldn't rise above. He's choosing to be a father to his three boys. He's choosing to not let his pain lead his actions. He's choosing to focus on building a future rather than grieving the past. He is pushing for growth and trying every day to stay on track. And I don't think we can ask any more than that from the people we care about.

My sister Hannah is a counselor and dedicates her life to helping those who struggle silently. My mom works for MD Anderson and is responsible for programs that span two separate locations. My sister Gi is in the Air Force and plans to be stationed overseas for her next assignment. My family is thriving, each of them making an impact in their own way.

I'm officially retired from the military, Master Sergeant, out with honor. That chapter closed with scars, stories, and a debt of gratitude I'll carry forever. I gave it everything I had, and the Army gave me more than rank or ribbons. It gave me perspective. A sharper lens. A deeper appreciation for what matters. Now, I live with intention. My life is still built around service and strength, but in new, more sustainable ways.

Today, I own rental properties across Georgia. They aren't just investments, they're anchors. Proof that stability is possible. That I don't have to rely on anyone else to build a future for my family. I can create that with my own hands, my own mind, and my own name.

My wife, my teammate in every sense, has built something of her own too. She's one of the top real estate agents in her region. Sharp, determined, full of heart. She didn't inherit success, but she earned every inch of it. I've watched her build, rise, and lead, and it's one of the most beautiful parts of our story. She's building her legacy, and I'm proud just to witness it.

Alongside her, I've carved out my space in coaching, bodybuilding, and powerlifting, and most recently, youth baseball for my son's team. Helping others discover strength they didn't know they had. For me, training isn't about aesthetics or ego. It's about transformation. About healing through discipline. About turning pain into purpose and rep after rep, learning how to stand taller in life, not just under the bar.

Eventually, we made yet another big move, one I didn't see coming until I could see that I had the freedom of choice. I no longer had to pick the best option based on where the Army told me to be. Now I could just seek the best location for my family and me to continue to grow and adapt to the new version of life that we found ourselves in. We relocated to Texas to be closer to family, my mom and her husband, Troy. He's not my biological father, but he's the only man I've ever truly called "Dad." He chose to be that. And that means everything. He's also an incredible grandfather and having him near my kids is something I'll never take for granted.

Texas gave us roots we didn't know we needed. My kids now grow up with grandparents close by, family dinners, more hugs, more presence. That sense of belonging is something I never had as a kid. And now they do. Everything did not come back together perfectly though. My relationship with my oldest son, Ezekiel, has been strained. The truth is, we grew apart. I spent too many years moving from state to state, warzone to warzone, duty station to duty station. I missed more than I should've. And when you miss enough, time starts to carve distance you didn't mean to create.

In his early years, I thought being a good example from afar was better than being flawed up close. I believed that going to war for something greater was somehow more noble than staying home. I know now, that was wrong.

He deserved presence, not just protection, and I carry that. There are things I could've done better. Things I can't undo. But I'll never stop trying. Zeke is a loving young man, and I am proud of him. As I am proud of all my sons, Ezekiel, Tristan, and Quinn. They are the best parts of me. And they will always deserve more than I had to give at their age. I know there may come a time when they speak their truth, when they look me in the eye and tell me what I got wrong. And when that time comes, I'll be ready. Not to defend myself, but to listen. To grow. To keep becoming the father they deserve. In the meantime, my door stays open. My arms stay wide. Zeke will always have a home with me. Because family doesn't expire. And love doesn't stop showing up. This chapter of life is quieter in some ways, but it's also louder where it counts. I'm not dodging bullets anymore, but I'm still in a fight. A fight for peace. For purpose. For a legacy that outlives me. I'm not building armor anymore; I'm building the castle. One brick at a time. But I didn't get here overnight. I got here one rep at a time. One hard conversation. One risk. One moment of saying, "I'm not done yet." This isn't the end of my story. But it is the beginning of a life I will hopefully learn to feel at peace living. I'm not just surviving anymore. I'm home. I'm present. I'm building something that matters. God put me on this path, so I will walk it until he gives me permission to take a final rest.

In Memory Of

The ones who never made it home, mentally, or physically.
This book is in memory to the warriors who gave
everything.
Not just their lives, but their laughter, their stories, their
futures.
To the ones who stood beside me in the dirt, in the heat,
in the silence, and chose courage.
You were more than uniforms.

More than names etched in stone.
You were brothers, mentors, jokers, fathers, and sons. And you mattered, deeply.
There is not a single page of this book that wasn't shaped by your sacrifice.
Not a single lesson I carry that wasn't paid for in blood, sweat, or memory.
This isn't just a story of survival.
It is a tribute to those who didn't get to write theirs.
Your legacy lives in how we live now.
In how we parent, lead, speak, and remember.
I carry your names in my heart,
your stories in my bones,
and your memory in every step forward I take.
Rest easy, brothers. We've got it from here.